ROUTLEDGE LIBRARY EDITI
ECONOMIC GEOGRAPHY

Volume 12

# THE INDUSTRIAL STRUCTURE OF AMERICAN CITIES

# THE INDUSTRIAL STRUCTURE OF AMERICAN CITIES
## A Geographic Study of Urban Economy in the United States

GUNNAR ALEXANDERSSON

Routledge
Taylor & Francis Group

LONDON AND NEW YORK

First published in 1956

This edition first published in 2015
by Routledge
2 Park Square, Milton Park, Abingdon, Oxon, OX14 4RN

and by Routledge
711 Third Avenue, New York, NY 10017

*Routledge is an imprint of the Taylor & Francis Group, an informa business*

© 1956 Gunnar Alexandersson

*British Library Cataloguing in Publication Data*
A catalogue record for this book is available from the British Library

ISBN: 978-1-138-85764-3 (Set)
eISBN: 978-1-315-71580-3 (Set)
ISBN: 978-1-138-88684-1 (Volume 12)
eISBN: 978-1-315-71455-4 (Volume 12)
Pb ISBN: 978-1-138-88685-8 (Volume 12)

**Publisher's Note**
The publisher has gone to great lengths to ensure the quality of this reprint but points out that some imperfections in the original copies may be apparent.

**Disclaimer**
The publisher has made every effort to trace copyright holders and would welcome correspondence from those they have been unable to trace. Due to restrictions of reproduction, Map 1 is not included with this book. You can access it at http://www.routledge.com/books/details/9781138886841/

# The Industrial Structure

# of American Cities

A GEOGRAPHIC STUDY OF URBAN ECONOMY

IN THE UNITED STATES

by Gunnar Alexandersson

ALMQVIST & WIKSELL · STOCKHOLM

Printed in Sweden by ALMQVIST & WIKSELLS BOKTRYCKERI AB · UPPSALA 1956

# Preface and Acknowledgments

THE METHODS presented in this book were tentatively worked out in a study on German, Scandinavian and Swiss prewar cities in 1948–51. This was when the author worked as research assistant to Prof. W. William-Olsson on his Economic Map of Europe. The difficulties of the European urban study were great, chiefly stemming from the heterogeneous statistical material. In 1952 the author lived a year in the United States on a research grant from the Stockholm School of Economics. He decided to apply his methods of urban industrial analysis to that country to test them with a larger statistical base and homogeneous statistics.

It has been possible to carry out this research only with the help of several persons and institutions in the United States and Sweden. Especially four persons should be mentioned: Prof. W. William-Olsson of the Department of Geography, Stockholm School of Economics, whose extensive research in the field of urban geography over a long series of years has inspired this book and who has always encouraged the author with helpful suggestions and criticism; Prof. Leslie Hewes of the Department of Geography, The University of Nebraska, who gave the research project all possible support while the author was visiting professor at his department in 1953/54; Prof. Ivar Högbom, Director of the Stockholm School of Economics and Head of the Department of Geography, who has followed the study with great interest and who has in several ways facilitated the research work; and last but not least the author's wife, Ingrid, whose encouragement and patience have been indispensable.

The author also wishes to express deep appreciation to Miss Emily Schossberger, Editor of the University of Nebraska Press, who has had an unusually heavy burden with the copy editing, English not being the mother tongue of the author.

For financial aid for the research work and the printing, the author is indebted to the Social Science Research Council of the Swedish Government.

*Maps and Diagrams.* Maps 2–16 and Fig. 1 were drawn at the Department of Geography, The University of Nebraska, by Mr. William B. Baker, Mr. G. Loyd Collier, Mr. Vincent E. Larocco, Mr. Morris L. Lewis, Mr. Richard C. Montgomery and Mr. James E. Thompson. Figs. 2–40 were drawn by Miss Britta Trulsson at the Department of Geography, Stockholm School of Economics and Map 1 by the Generalstabens Litografiska Anstalt, Stockholm. For the map work the author has drawn much inspiration from the Atlas of Sweden and from discussions with its editor Dr. Magnus Lundqvist and his assistant Mr. Olof Hedbom. Baker, Collier, Larocco, Lewis, Montgomery, Thompson and Miss Trulsson have made several valuable suggestions: the maps are the result of a team work. Unfortunately the printed two-color maps do not render justice to the originals which were drawn on a much larger scale for exhibition purposes. Because of prohibitive costs they could not be redrawn for the scale in which they appear in the book.

In writing the text the author has drawn heavily upon articles and books in geography and related subjects. It has not been possible to indicate all sources in the footnotes. Prof. W. William-Olsson, Stockholm, Prof. Leslie Hewes, Lincoln and Prof. John E. Christensen, Houghton, Mich., have critically read the manuscript and parts have also been read by Prof. Ivar Högbom, Prof. Gunnar Arpi, Dr. Olov Hölcke and Dr. Roland Artle, Stockholm and Mr. Bert Rudolf, Karlsruhe, Germany. The author also wants to acknowledge valuable suggestions made by his colleagues of Lincoln at a preliminary stage of the work, Prof. Esther S. Anderson, Prof. Nels A. Bengtson and Prof. Robert G. Bowman, and help in other form from Dr. Robert C. Klove, Assistant Chief, Geography Division, Bureau of the Census.

*Department of Geography*
*Stockholm School of Economics*
*December 1955*

GUNNAR ALEXANDERSSON

# Contents

# Introduction

## Purpose of the Study

In this book an attempt is made to analyze the distribution of the urban population in an industrialized country. The United States was chosen as an object of study because it has the largest population for which homogeneous and comparable statistics are available.

The technique of mapping population distribution has been well developed by geographers. The logical next step would be to analyze the population map, but few geographic studies can be seen as attempts in this direction. On the whole, geographers have neglected the study of the urban industries, which employ over four-fifths of the gainfully employed population in the United States. This is reflected by textbooks in economic geography, which generally emphasize very strongly the rural industries and mineral production, a rather small fraction of the total economy in industrialized countries. Geographers often focus attention on the surface of land in the cultural landscape, not on man, the active agent. On most distribution maps, to take one example, the city of New York has the weight of the land area it covers, not one which corresponds to the role of the city in the matter under study.

The first step in the quantitative analysis of population distribution, according to the method suggested here, is the breaking up of the total population into its components:
the industries in which people earn their living (Fig. 1). If comparable maps are made for these industries it should be possible to answer the first question which is asked when we are confronted with a population map: Why is the population distributed in this way? Or rather: the question is pushed one step backward. Why is the textile industry, the steel industry, or retail trade located where it is? This raises the general problem of industrial location, which has attracted much attention from geographers and economists. A standardized cartographic presentation of all industries will automatically draw attention to the interdependence of different forms of economic activity, which may not be given proper consideration in studies of single industries.

The scope of the present study prevents a detailed analysis of each industry. The discussion of individual industries should be seen as an attempt to summarize some pertinent historic developments, some locational discussions presented in the literature, etc. The reader should not expect to find much new information in these brief comments, which chiefly are intended as a guide when he wants to interpret the maps. The scientific contribution of the study lies in the methods of arranging and presenting the statistics.

The manufacturing industries are presented on a series of maps (2–16) made according to an absolute method. These maps should give a reasonably true impression of the regional differentiation within the manufacturing industries. The functional differentiation of manufacturing in cities of 10,000 or more inhabitants can be grasped from a comparative study of a series of relative maps and from Appendix 1.

The service industries, engaging more than half of the gainfully employed population in the United States, are ubiquitous in American cities. Maps of retail trade and other service industries made according to an absolute method would be of limited value. They would give a visual impression of a distribution almost equal to that of the total urban population. For this reason, only relative maps have been made for the service industries. The significant regional and functional differentiation in these industries is brought out on the relative maps, made according to the same principles as those for the manufacturing industries, and in correlation tables in the text.

The relative maps make a study of the regional differentiation in all urban industries possible and meaningful even if the suggested method for determining the ratio of city forming and city serving industries should prove to be unsatisfactory. The latter hypothesis is basic, however, to the comparison of different industries and to the study of the functional differentiation. To make comparisons easier when the interest is focused on individual cities, the most important city forming industries have been listed in an appendix for all cities with more than 10,000 inhabitants (App. 1). This appendix will also be useful for the interpretation of Map 1, which shows the relative importance of manufacturing and service industries in American cities.

As the differentiation within agriculture —the predominant rural industry—is well covered by maps continuously prepared by official agencies, it has seemed reasonable to concentrate this study on the urban industries.

The present study was made without any thought to its practical usefulness. The author can, however, imagine several instances in which it may be of use in applied research. Some random examples are listed below.

The social planner dealing with small areas (cities, counties) often feels a need for comparison with other areas. In an isolated city study it is convenient to make comparisons with averages for the nation or the state, which are easily computed, but it should be of value to see the city in its functional as well as in its regional setting. In many instances it may be more meaningful to choose cities with similar industrial structures as controls.

If the problem is to attract new industries to a city it should be valuable to know the national distribution patterns of these industries and the locational factors that have been instrumental in the shaping of these patterns. It should be of interest to know how the industries in question are associated with other industries in cities of different sizes, functions and geographic locations.

When a population prognosis is made for a city, attention should be concentrated on the city forming industries and their prospects. This can easily be done for a one-sided town, but it may be difficult to determine the relative importance of the city forming industries for cities with a complex industrial structure.

Last but not least: the author believes that maps on American industrial distribution patterns will facilitate the understanding of similar patterns in Europe. How much are the European patterns influenced by the fact that Europe is split into small national markets? Compare, for instance, the distribution of automobile manufacturing in the United States and in Europe: the American industry concentrated to a relatively small area centered around Detroit, the European industry scattered over a large number of countries. Or the textile industry: in the United States a concentration to an almost continuous textile district stretching from New England to the southern Appalachian Piedmont, with practically no textile industry in most of the other states; in Europe one or more textile districts in every country. What would be the likely long-run effect on various distribution patterns if some or all of the European countries formed a customs union? This and other questions will present themselves if we study comparable maps of industrial distribution patterns for the United States and Europe.

## Rural and Urban Industries

From the geographic point of view there is a basic difference between farming, forestry and fisheries on the one hand and the remaining industries on the other. The former demand large surfaces for their production; the latter have limited areal requirements. This distinction is fundamental to an understanding of the population distribution of an area.

As there is a definite limit to the distance between home and working place which can conveniently and economically be covered every day, farm population tends to be spread out on the land, either in scattered farmsteads, as is usually the case in the United States, or in small villages, as

is probably most common in the rest of the world. Only in areas with a monoculture, requiring the farmers' attention during a relatively short period of the year, is it possible for the farm population to allow themselves a longer commuting distance and thus to live in larger agglomerations and enjoy the social and economic benefits of such a life.

In the wheat and wine producing areas of southern Italy and Spain such agglomerations will reach a size of 40,000 inhabitants. There the majority of the population is engaged in farming, and the agglomerations are villages. In the United States, Kollmorgen and Jenks have called attention to a recent development in the wheat areas of the Great Plains.[1] It is rather common there that farmers live in towns of, say, 2,000 inhabitants, where they often have an additional occupation. They are "sidewalk farmers". The farm population, however, is a rather small minority in these towns. The agglomeration is still functionally a town, not a village.[2]

These examples are exceptions. As a rule the farm population must live close to the producing surfaces, either on scattered farmsteads or in small agglomerations. In the United States there are no agglomerations with 10,000 or more inhabitants in which the farm population dominates or even constitutes a considerable percentage. Usually their share is only one or possibly a few per cent. Most of these people live in the fringe area where town and country meet. Economically they do not belong to the city; they are a relict. Their production surfaces will soon be swallowed by the growing city, and other farmers in the new fringe will be counted by the census as city people. The variations in the relative number of farmers in American cities of the size which is dealt with here will reflect more than anything else the arbitrariness in

1. W. M. Kollmorgen and G. F. Jenks, "A Geographic Study of Population and Settlement Changes in Sherman County, Kansas," *Transactions of the Kansas Academy of Science* (1951).

2. A few live well away from their land in larger cities and are "suit-case farmers," staying on the land only for the planting and the harvest.

the drawing of city limits. These variations are of no geographic significance and the small farm population of American cities can very well be disregarded in an urban study.

The production conditions of forestry and fisheries are similar to those of farming. These industries are also tied to production surfaces. Their population distribution is different, however. The fishermen's homes are usually on land. Their product will often be brought to a big port, but the fishermen themselves need not live there. They often live in small agglomerations along the coast, usually between the landing port and the most frequented fishing grounds.[1]

Fig. 1 shows that one-third of the people engaged in forestry and fisheries in the United States live in towns of more than 2,500 inhabitants, and over half of them live in smaller towns and are thus counted by the census as rural. Their total number is, however, so small—much smaller than that of any other American industry, less than 0.25 per cent of the total employed population—that a special investigation to determine their population distribution has not been justified in this study.[2] They have been treated in a conventional way and have been grouped together with farming under the heading of *rural industries*.

The rural industries give employment only to a minority of the gainfully employed population in most Western countries of today (12.4 per cent in the United States 1950). The majority live in urban agglomerations, employed in manufacturing and service industries. As these two large groups of industries are to a very high degree concentrated in cities, they will, in accordance with common usage, be referred to as *urban industries*.[3]

The urbanization and the increase—both absolute and relative—of manufacturing and of service industries during the last hundred years have gone hand in hand. They are two expressions of the same phenomenon, one referring to the changed settlement pattern, the other to the changes in production which have caused the new population distribution. To agglomerate production geographically must be of definite advantage within urban industries, since we know empirically that such a concentration always takes place. The fact that urban industries have limited areal requirements does not imply that they must cluster in certain spots and not, like the rural industries, be spread out rather evenly over the land surface.

There must be positive factors bringing about this concentration. In the terminology of the economists, these factors can be

1. Gloucester, Massachusetts (25,000 inhabitants), is the city in the United States which is most influenced by fisheries. Even in this exceptional case the fishermen do not make up more than 15 per cent of the gainfully employed population.

2. Logging is included in the manufacturing of furniture, and lumber and wood products. Often logging is an integrated part of the saw mill operation. Seen separately, however, it is a form of areal production (see note 3 below), and should logically belong to the rural industries.

3. Fig. 1 shows that 75–85 per cent of the gainfully employed people in most urban industries live in towns with more than 2,500 inhabitants. Of the remainder the majority live in smaller towns. A detailed study of the maps of manufacturing industries (Maps 2–16), reveals that the dots often are concentrated close to the centers of the red circles indicating that a considerable share of the rural population within these industries may be rural only in a formal sense. With present means of transportation it is possible to live far out in the countryside or in a small town and still work in the city.

The terms rural and urban industries are here used as synonyms for areal and local production, often utilized by Scandinavian geographers. A country school teacher, to take one example, is employed in an urban industry, but she belongs to the rural population since she lives in a rural area. A farmer living inside a city limit belongs to a rural industry but to the urban population. In a less urbanized country than the United States a majority of the people employed in an "urban industry" may live and work in purely rural areas. The terms areal and local industries are therefore theoretically preferable, but since they may sound unfamiliar to the American reader they have not been used here.

grouped together as "advantages of specialization and division of labor." On the side
of production—disregarding temporarily the
distribution of the final product and the
procuring of raw materials—highest efficiency lies in geographic concentration.
The advantages of concentration are, however, counterbalanced by the resistance offered the raw materials and the final product
by the distance factor. Out of this balance
come distribution patterns which are quite
different from industry to industry. They
are dynamic in character and will change
with technical innovations affecting one
side of the balance or the other. To describe
and interpret these patterns is a central task
in economic geographic research. To study
their interaction in the production nodes,
the cities, is an essential part of urban
geography.

It can be expected, then, that the agricultural population, producer of raw materials for urban industries and consumer of
urban products, will have an influence on
the distribution of urban population which
is out of proportion to its own quantity.
Other raw material producing and procuring
industries, especially mining and shipping,
will have a similar effect. The industries
attracted to a city or a region by its port
location or its coal mining will often employ
many more people than are engaged in the
mere handling of goods or mining of coal.

The geographic concentration of the urban industries to production nodes does not
necessarily mean that production is carried
on within one or a few big plants in each
town, but can also, as in apparel manufacturing or retail trade, be brought about
by small establishments flocking to the
same locality.

1. Construction and mining are in this study included in manufacturing or the production of goods.
All other urban industries are referred to as service
industries, they produce services.
2. Ubiquitous with reference to cities.

## Sporadic and Ubiquitous Industries

The general distribution pattern of the
urban industries is very different from case
to case. Of the manufacturing industries
only a few are represented in all cities:
construction,[1] printing, and food manufacturing. No city lacks building contractors,
a newspaper, and a few food plants. But the
distribution of other manufacturing industries in the United States reveals a sporadic
occurrence, they are *sporadic industries*. For
each of these it is characteristic that the
industry is not at all represented in many
cities but plays an important or dominant
role in the economy of others.

On the other hand, all service industries,
as well as the three above-mentioned manufacturing industries, are to be found in all
cities, and they may appropriately be la-
belled *ubiquitous industries*.[2]

Every city has retail and wholesale trade,
transportation facilities of different types,
schools, hospitals, hotels, recreation establishments, etc. In the ubiquitous industries
large-scale production for a regional, national, or international market is as a rule
of subordinate importance in comparison
with production for the city itself and its
immediate trade area. There are of course
exceptions. Some cities are dominated by
institutions of higher education, military
garrisons and other government establishments, big hospitals, and so on. Some of
the biggest cities provide wholesale trade,
banking, and other services for a national
and international market.

The ubiquitousness of the service industries and a few of the manufacturing
industries does not mean that they are of
equal importance in all cities. There are
significant variations between cities of different size and different functional type as
well as regional variations. Cities with more
than a hundred thousand inhabitants thus

have an above-average share of people employed in wholesale trade, but less than average employment in retail trade. Extreme manufacturing cities have a low percentage of people employed in most or all service industries, whereas a city with relatively many people engaged in retail trade usually is well equipped also with other service industries. Cities of the South have a considerably above-average percentage of people employed in private households, reflecting a still existing difference in the way of life between this area and the rest of the country. Construction employs more people in the cities of the Southwest than in those of other areas.

The degree to which various industries are ubiquitous or sporadic in their occurrence can be judged from the relative maps, but especially from the distribution diagrams on these maps. The diagrams were constructed in the following way. For every city the ratios of the 36 urban industries were calculated and expressed in per cent of its gainfully employed "urban" population. The small farm population living within the city limits was excluded. For every industry the cities were arranged according to their rates. In the distribution tables thus obtained, a division was made into ten equal groups or decils. The diagrams were based on these tables and they show that the industries are of two main types:

1. Those with a very low (below 0.2 per cent) ratio in many cities (10 to 70 per cent of the total number) but with very high values (above 20 per cent) for a few cities. The manufacturing industries belong to this type with the exception of construction, food processing and printing, which are of the second type.

2. Those with some ratio of the industry for all cities and with relatively small differences between cities in the first and the tenth decils. The service industries belong to this group. Railroads, medical services, education and public administration are exceptions: some cities show very high values (above 20 per cent). In this respect they remind one of the sporadic industries in the first group, but they are at the same time, and in their main characteristic, ubiquitous.

## City Forming and City Serving Production

Within all human groups there is division of labor and specialization associated with an exchange of goods and services. In the family group such a division of labor has occurred in all times and in all cultures. Even between groups an exchange of goods and services has always taken place, indicating also a division of labor. This exchange has reached its fullest development in the modern Western world, where it is now possible for a man to satisfy a varied need for consumption of goods and services in spite of the fact that he may devote all his productive time to such specialized tasks as teaching other people's children or producing details of an automobile motor.

It can be assumed *a priori* that the exchange between groups will be relatively larger the smaller the groups are. Within a very large group it is possible to obtain a high degree of satisfaction of wants for the individuals even with a limited participation in the intergroupal exchange. A division of labor and a specialization can be arranged within the group. Large countries, other things being equal, will have a smaller percapita foreign trade than small ones. It is not a pure coincidence that a large country like the United States, not to mention the Soviet Union, is more protectionist than small countries like Belgium or Sweden.

The big country can better afford a protectionist policy because it will not lower the optimum standard of living as much as in the small country.

Cities,[1] considered as economic units, are much more dependent than countries on the exchange of goods and services with other groups. A larger per cent of their population can be expected to be engaged in production for the outside world.

As the city is not an autonomous economic unit there has been no practical need for a distinction between production for the city

people and production for the outside world. That there nevertheless is a great interest in such a distinction is evident from the many suggestions of a terminology which have been put forward.[2] An analysis of the industrial structure of a city in order to determine its economic functions can hardly do without this distinction.

In the present study production for the city's own inhabitants is referred to as *city serving production*. Types are grade schools, small bakeries, neighborhood retail stores. Both manufacturing and service industries can be city serving.

The main attention in this study will, however, be directed towards industries which produce for a market outside the geographic city limit.[3] They are the agglomerative element, the *raison d'être* of the city, and might therefore be termed *city forming industries*. They bring money to the city, which is used to pay for the imports of such goods and services in which the city is deficient.

It is possible to imagine a large self-sufficient agglomeration in which all industries are city serving. In a primitive, feudal society it would be possible to have a large group of people living together in a city-like agglomeration, growing their food and other raw materials on the surrounding land and trading mostly with each other and very little with the outside world. Agglomerations that come rather close to this type exist in southern Italy, in Spain and in other parts of the world.[4] Northwest European towns of the period before the Industrial Revolution were also self-sufficient to a degree that is now unknown there. Many citizens were part-time farmers. Much work that now, at least partly, has been taken over by city forming industries was carried on in the homes.[5] But neither the present-day agricultural agglomerations of southern Europe nor the northwest

1. The terms *city* and *town* are used in this study as synonyms.

2. Here only a few shall be mentioned:
Ekstedt suggested *primary city forming industries* (primärt stadsbildande branscher) and *secondary city forming industries* (sekundärt stadsbildande branscher). H. W:son Ahlmann, I. Ekstedt, G. Jonsson, W. William-Olsson, *Stockholms inre differentiering* (Stockholm, 1934), p. 40.
William-Olsson made a distinction between *exchange production* (bytesproduktion) and *self production* (egenproduktion). Neither Ekstedt nor William-Olsson attempted a quantitive determination of these entities. W. William-Olsson, *Stockholms framtida utveckling* (Stockholm, 1941), p. 12 ff.
Homer Hoyt and others have distinguished *basic* and *non-basic* industries and various ways of measuring them have also been suggested. For exhaustive accounts of the American discussion of the economic bases of cities, see Richard B. Andrews, "Mechanics of the Urban Economic Base," *Land Economics* (1953) and John W. Alexander, "The Basic-Nonbasic Concept of Urban Economic Functions", *Economic Geography* (1954).
In a study of the Dutch city of Amersfoort the terms *primary* (primair) and *supporting* (verzorgend) industries were used and a method was also worked out to estimate their size, page 16. L. H. Klaassen, D. H. van Dongen Torman, L. M. Koyck, *Hoofdlijnen van de sociaaleconomische ontwikkeling der gemeente Amersfoort van 1900–1970* (Leiden, 1949).

3. Note the difference between the administrative city and the (usually larger) geographic city. These terms are further discussed on page 23.

4. They are conspicuously large, say, 10 to 40 thousand inhabitants, mostly in areas dominated by one or a few crops, often wheat or wine, requiring the farmers attention for the work in the most distant fields during a relatively short period of the year. Both wheat and wine are staple food in these areas and a very large share of the crop is consumed locally.

5. For example: many food products that were prepared at home in the old days, are now furnished in finished or semi-finished form by the food manufacturing industry, to a large extent from out-of-town plants.

European towns of a hundred or more years ago are urban in the modern northwest European and American sense. In America and northwestern Europe of today, cities are highly specialized nodes of production with a very big exchange of goods and services with other urban places and with rural districts.

It is relatively simple to estimate, at least approximately, how large a share of a country's national product goes to the outside world in exchange for imported goods and services. Statistics on foreign trade are available for most countries, and it is usually possible to determine the amount of exports of services such as shipping, tourism, etc. Since internal trade is inadequately covered by statistics, to get an idea of the relative importance of city forming and city serving industries, we must resort to indirect methods.[1] Data on "exports" and "total product" of a given city, which would allow a direct approach, are not available.

An interesting attempt to single out and determine quantitatively the city forming industries is presented in a study of the Dutch city of Amersfoort. For the eighteen biggest cities in the Netherlands the percentage of people employed in different manufacturing industries is figured. It is assumed that the city with the minimum equipment of a given industry barely has city serving (verzorgend) production. The other cities with a smaller or larger "surplus" over this rate, in addition to their city serving production, also have a city forming (primair) portion of the industry in question. The sizes of the two components are determined by the surplus.

Other methods for measuring the proportion between city serving and city forming production have been suggested. Hoyt and others compare the ratio of a certain industry in a given city with the ratio for the country as a whole, the national

average. This is convenient in an isolated urban study, as both ratios are easily computed. Theoretically, however, it seems to be less satisfactory. The national average is a statistical abstraction, difficult to interpret. This is especially true for the sporadic industries. The national average of automobile manufacturing in the United States, to take one example, is 1.5 per cent, but only 12 per cent of the cities have such a high ratio of people employed in the automotive industry. Seventy per cent of the cities have a rate of 0.2 per cent or lower.

In Detroit 28 per cent of the gainfully employed population is engaged in automobile manufacturing. One and a half per cent are, according to Hoyt, needed to supply the city's own population with a normal amount of cars. The 1.5 per cent represents the city serving (non-basic) portion, the remaining 26.5 per cent is the city forming (basic) ratio of automobile manufacturing in Detroit.

The national average is evidently influenced by such irrelevant facts—for the individual city—as foreign trade and the

---

1. The research staff of *Fortune* magazine (Oskaloosa, Iowa), Forbat (Skövde, Sweden) and Alexander (Oshkosh and Madison, Wisconsin) attempted the direct method by asking the firms of a city what shares of their production were for the city's own population. Interesting as these attempts are, they cannot be applied in a comparative study of cities. It is also doubtful if they will yield safer results. It is for instance hard to believe that the average retail establishment can judge within reasonable margins of error how much it sells to customers residing in the city, as no records are kept of cash sales. The average business man may not know the exact location of the city limit, much less the exact residence of his customers.
"Oskaloosa vs. the United States" *Fortune* (April, 1938).
Forbat, "Utvecklingsprognos för en medelstor stad. En studie över näringsliv, befolkning och bostäder i Skövde." *Statens kommitté för byggnadsforskning, rapport nr 18* (Stockholm, 1949).
John W. Alexander, "The Basic-Nonbasic Concept of Urban Economic Functions," *Economic Geography* (1954) summarizes the results of the Oshkosh and Madison studies, published by the Bureau of Business Research, School of Commerce, University of Wisconsin (1951 and 1953, respectively).

configuration of the national boundary. If Mexico joined the United States all national averages would change. Or if the United States were divided into autonomous countries, the national average of automobile manufacturing with which Detroit would then be compared would be quite different from the above ratio. These two hypothetic changes would, everything else being equal, have no direct effects on the industrial structure of Detroit, but they would definitely influence the city serving ratio computed according to Hoyt.

The present study of 864 American cities with 10,000 or more inhabitants has been based on the same theory as the Amersfoort investigation. Accordingly, the whole of Detroit's automobile manufacturing should be considered as a city forming industry. Some of the cars made in Detroit will undoubtedly be sold to people living in the city. But all automobile factories in Detroit were built for a bigger market than the city itself offers. This is in contrast to laundries and grade schools, which, even if they occasionally may have an out-of-town customer, were built for the city market or perhaps for only a section of it.

All sporadic industries are of the same type as automobile manufacturing, whereas the ubiquitous industries in general are more complex than the type represented by laundries and grade schools. Many ubiquitous industries, like retail shopping stores, wholesale establishments, etc., are in most cities based on both the city market

and a wider market, usually the trade area of the city. They are partly city serving, partly city forming. In some cases, illustrated by insurance companies, banks, and public administration, the industries have both a sporadic and a ubiquitous component, reflecting the hierarchic structure of all large organizations. There can also be a sporadic component in a generally ubiquitous industry for other reasons. Movie production, the armed forces, and the universities are examples of such sporadic branches within ubiquitous industries,[1] whose distribution patterns are shaped by conspicuous advantages of large-scale production similar to those found in manufacturing. Most ubiquitous industries are both city forming and city serving. Our problem is to determine the ratios of the two components.

The method employed here to determine the ratio of city serving production will apparently give lower values than methods used by Hoyt and others.[2] Our values answer the question: *what ratios in different industries are a necessary minimum to supply a city's own population with goods and services of the type which are produced in every normal city?*

The following procedure was used to obtain these ratios. For every industry two points on the cumulative distribution diagrams (page 14) were tentatively chosen, one and five per cent from the origin respectively, which represent cities number 9 and 43, as there are 864 cities all together. These values and not the very lowest ones were chosen to avoid extreme ratios representing such agglomerations as Midway-Hardwick, Georgia, and Kings Park, New York, which are just large hospitals with some settlement around them and are not towns in the ordinary sense. They reach the size of 10,000 inhabitants only because their patients are included. Other agglomerations with very low rates are the two newly built

---

1. Entertainment, public administration and education respectively.
2. It is of course perfectly possible to imagine a (large) city with the same industrial structure as the country (with the exception for agriculture), which would give a city serving percentage of close to 100. Comparisons with the national averages will give by far the highest values for city serving production; the methods employed by Forbat and Alexander should theoretically give somewhat higher values than the method employed in the present study.

"atom cities" Oak Ridge, Tennessee, and Richland, Washington. They can, however, hardly be considered as "normal" yet. Their inhabitants probably go to nearby cities for services (refers to 1950) which normally would be available in agglomerations of their size and which will probably be available when they have reached a higher degree of maturity.[1]

The values for the two tentative points were tested against actual city structures to find if there are any cities corresponding to the two structure models. It could *a priori* be expected that such barely "self-sufficient" cities might be found among extreme manufacturing cities or one-sided service towns in regions with a high city density. They would have a very small trade area, small enough to be considered as a negligible quantity as people in the neighborhood would prefer to trade in better equipped cities. On the other side they would have all the service production which normally is available in any city of 10,000 or more inhabitants for its own population. It should be self-sufficient but not deficient in such services.

Among agglomerations which resemble the first tentative model are Kannapolis, N.C., and Bristol, R.I. The first one is a company-owned textile mill town with mills established in 1877,[2] about 22 miles northeast of Charlotte, in a region with an agglomeration density among the highest to be found anywhere in the United States. It is an extremely one-sided textile town with 28,400 inhabitants. The second one, dominated by the manufacturing of saddles, rubber products and textiles, is located just 12 miles southeast of Providence. After considering the character and location of these agglomerations they have been judged to be deficient in essential service industries. According to this structure model, city serving production employs 28.3 per cent

of the gainfully employed population of an American city.

If the norm structure is built instead on values 5 per cent from the origin on the cumulative distribution diagrams, city serving production employs 37.7 per cent. The curves of the ubiquitous industries have all leveled off at this point; the extremely low values are all to the left. It therefore seems that a theoretic minimum structure may be based on these values. Among cities with an industrial structure very similar to that of the second tentative norm structure are Woonsocket, R.I., Tamaqua, Pa., and Thomasville, N.C. Woonsocket is a textile town with 40,100 inhabitants. Tamaqua is a mining town with 11,500 inhabitants in the Pennsylvanian anthracite region. It is also a railroad and chemical town of B-type. Thomasville is a furniture and textile town with 11,200 inhabitants. All three cities are located in regions with a high density of urban agglomerations.

Local studies of these or similar agglomerations need to be undertaken for a safer testing of the norm structure.

It is evident that even if the values found are valid for cities of 10 to 50 thousand inhabitants, which make up the bulk of the total number in this study, they may not be applicable to larger cities. In New York with its 13 million inhabitants, there will be a relatively larger exchange within the city than is possible within a smaller agglomeration. With increasing size of the town, city serving production can be expected to increase in relative importance,

1. An indication in this direction is the fact that the small twin-cities of Pasco and Kennewick near Richland, which themselves have had their population increased with about 300 per cent to 20,300 inhabitants in the 1940's, have high ratios for "Other retail trade" (12.9 %), "Eating and drinking places" (5.1 %), etc. The pronounced service centers of Walla Walla and Yakima are within reach for less frequent shopping for the inhabitants of Richland.

2. Columbia Lippincott Gazetteer (New York, 1953).

TABLE I. Industrial Structure of Some Towns Mentioned in the Text, Pages 17-18.

| Industry | $k_1$ | $k$ | Kings Park | Rich-land | Kanna-polis | Bristol | Woon-socket | Tamaqua | Thomas-ville |
|---|---|---|---|---|---|---|---|---|---|
| Mining | 0 | 0 | 0 | 0 | 0 | 0 | 0 | 279 A | 0 |
| Construction | 26 | 35 | 31 | 98 C | 35 | 46 | 41 | 29 | 41 |
| Lumb. & Furn. | 0 | 0 | 0 | 0 | 1 | 1 | 3 | 0 | 295 A |
| Prim. metal | 0 | 0 | 2 | 0 | 0 | 77 C | 8 | 7 | 1 |
| Fabr. metal | 0 | 0 | 1 | 0 | 1 | 3 | 2 | 2 | 1 |
| Machinery | 0 | 1 | 1 | 0 | 0 | 5 | 49 | 2 | 2 |
| El. machinery | 0 | 0 | 1 | 0 | 0 | 14 | 1 | 0 | 1 |
| Motor vehicles | 0 | 0 | 0 | 0 | 0 | 0 | 1 | 1 | 0 |
| Transp. equipm. | 0 | 0 | 11 | 0 | 0 | 2 | 0 | 0 | 1 |
| Oth. durable | 1 | 2 | 4 | 0 | 1 | 45 | 8 | 3 | 16 |
| Food | 3 | 7 | 3 | 2 | 5 | 9 | 8 | 18 | 8 |
| Textile | 0 | 0 | 0 | 0 | 695 A | 196 B | 465 A | 9 | 256 A |
| Apparel | 0 | 0 | 3 | 0 | 1 | 0 | 15 | 46 | 11 |
| Printing | 5 | 7 | 2 | 7 | 6 | 5 | 7 | 8 | 7 |
| Chemicals | 0 | 1 | 1 | 664 A | 0 | 1 | 3 | 115 B | 2 |
| Oth. nondurable | 0 | 1 | 1 | 0 | 0 | 298 A | 39 | 1 | 1 |
| Railroads | 2 | 4 | 6 | 2 | 2 | 2 | 3 | 127 B | 6 |
| Trucking | 3 | 5 | 2 | 0 | 9 | 6 | 9 | 8 | 4 |
| Oth. transport | 3 | 5 | 8 | 2 | 9 | 5 | 8 | 4 | 5 |
| Telecommunicat. | 4 | 6 | 0 | 2 | 3 | 7 | 8 | 8 | 5 |
| Utilities | 6 | 9 | 3 | 2 | 2 | 8 | 10 | 12 | 8 |
| Wholesale | 9 | 14 | 2 | 1 | 4 | 12 | 10 | 15 | 10 |
| Food retail | 23 | 27 | 26 | 14 | 28 | 28 | 29 | 33 | 25 |
| Eating places | 18 | 21 | 18 | 13 | 10 | 19 | 24 | 30 | 21 |
| Oth. retail | 63 | 80 | 38 | 37 | 65 | 58 | 80 | 77 | 78 |
| Finance | 12 | 18 | 12 | 7 | 12 | 17 | 19 | 18 | 15 |
| Business serv. | 1 | 2 | 2 | 5 | 1 | 2 | 4 | 2 | 3 |
| Repair serv. | 8 | 11 | 3 | 2 | 10 | 13 | 15 | 21 | 10 |
| Priv. households | 10 | 13 | 13 | 13 | 28 | 14 | 9 | 5 | 37 |
| Hotels | 2 | 3 | 6 | 4 | 4 | 3 | 2 | 9 | 2 |
| Oth. pers. serv. | 17 | 21 | 14 | 13 | 22 | 22 | 18 | 16 | 33 |
| Entertainment | 5 | 7 | 6 | 7 | 6 | 2 | 7 | 7 | 10 |
| Medical serv. | 13 | 18 | 736 A | 21 | 6 | 11 | 17 | 21 | 13 |
| Education | 22 | 26 | 22 | 43 | 16 | 28 | 22 | 25 | 27 |
| Oth. profess. serv. | 10 | 12 | 5 | 8 | 10 | 13 | 14 | 14 | 25 |
| Publ. administ. | 17 | 21 | 21 | 43 | 7 | 23 | 40 | 22 | 20 |
| | 283 | 377 | 1,000 | 1,000 | 1,000 | 1,000 | 1,000 | 1,000 | 1,000 |

$k$ = values representing point five per cent from origin on cumulative distribution diagrams (Figs. 2–40).
$k_1$ = same as above for point one per cent from origin.

not to decrease. When it is stated that 37.7 per cent of the gainfully employed popula-tion in American cities are engaged in city serving production, this is to be considered as a minimum figure relevant for small cities. In bigger urban agglomerations it can be expected to be higher.[1]

For retail trade the national average of

1. These logical deductions are supported by the results of Alexander's studies on Oshkosh and Madi-son, Wisconsin, both made with the same methods. For Oshkosh (41 thousand inhabitants) Alexander found a city serving (non-basic) percentage of 37.5

employment[1] should coincide with the *k*-value if this value has been wisely chosen and if there is an "American standard" in the sense that 1,000 average Americans, rural or urban, living in New York or in Mississippi, give employment to the same number of retail trade employees. This must be true as the imports and exports of retail trade services (by tourists, etc.) are infinitesimal in relation to the total retail trade of the United States.

The assumption of an American standard, implied in the present study, is of course an approximation. It is therefore not surprising that the national average for Other retail trade, 9.2 percent, differs from the *k*-value, 8.0 percent. There is a regional difference in the standard of living as well as a difference between rural and urban areas. In California with an above average urbanization and high standards of living, the per capita retail sales were 26 per cent higher than the United States average in 1948[2] and the employment in Other retail trade in 1950 was 17 per cent higher than the national average. These differences are mostly due to higher levels of income in California and only in some degree to the presence of a relatively large floating population of nonresident tourists and other visitors.[2]

The high percentage of retail trade employment for some states with high standards of living influence the national average but not the *k*-value. The case of retail trade raises the question if it would not have been motivated with different *k*-values for different regions. Private households is another industry in which there are at least two different "American standards", one for the South and one for the rest of the country.

The differences between national average and *k*-value for other ubiquitous industries are largely due to the inclusion of sporadic branches in these industries. Thus the national average for Education includes college employees, but these do not influence the *k*-value, as most cities do not have a college.

## The Functional Classification of Cities in Recent Literature

There seems to be a general agreement among students of urbanization that the industrial structure of cities holds a key to an understanding of their location and growth. In most urban monographs data is presented on the industrial structure. Many geographers and urban sociologists have made comparative studies of cities and suggested functional classifications, but these used to be of a qualitative rather than quantitative character.[3] City classification based on the industrial structure and stated statistical criteria is a recent development.[4]

and for the larger Madison (110 thousand inh.) a percentage of 45.1. Alexander, *op. cit*, p. 251.

In parenthesis it can be mentioned that Forbat found a city serving percentage of 45.3 for Skövde, Sweden (19,000 inhabitants), and *Fortune* 39.4 for Oskaloosa, Iowa (11,000 inhabitants).

1. Expressed in per cent of the total gainfully employed population, including the rural industries.

2. "Economic Survey of California." Reprint from *California Blue Book* (1950) p. 56.

3. Harris lists American works by Tower (1905), Anderson and Lindeman (1928), Carpenter (1931), Van Cleef (1937), Muntz (1938), and Gist and Halbert (1941). Chauncy D. Harris, "A Functional Classification of Cities in the United States," *Geographical Review* (1943).

Chabot in a small but illuminating volume on urban geography gives a brief survey of the history of the subject, mentioning pioneer works by Ratzel (1891), Meuriot (1897), Schlüter (1899), Hassert (1907), Clouzot (1909), Maunier (1910) and others. Georges Chabot, *Les Villes* (Paris, 1948).

Important contributions to the functional classification of cities have also been made by Aurousseau, Jefferson, Fawcett, Dickinson and others.

4. One reason why the pioneers in the field did not attempt to elaborate quantitative classification schemes is obvious. The census authorities had to await the arrival of modern business machines to be able to tabulate and publish at reasonable cost detailed statistical data for such small areas as cities of, say, five or ten thousand inhabitants. In the majority of countries such data are still not available. The stu-

The first two classifications of this type were published in 1943 by Chauncy D. Harris and W. William-Olsson.

Harris[1] classified the American cities of 10,000 or more inhabitants in eight groups: (1) Manufacturing cities (of two intensities), (2) Retail centers, (3) Diversified cities, (4) Wholesale centers, (5) Transportation centers, (6) Mining towns, (7) University towns, and (8) Resort and retirement towns.

William-Olsson in an official investigation on the industries of northern Sweden (Norrland)[2] classified the towns in three

---

dents of a few decades ago did not have sufficiently detailed statistics available to make quantitative classification schemes. Dr. Herman Hollerith, who invented one of the best known systems of recording and tabulating data, was a statistician in the Census Bureau at Washington. J. B. Walker, *The Epic of American Industry* (New York, 1949), p. 275.

1. Harris, *op. cit.*, p. 88.
2. "Utredning angående Norrlands näringsliv," *Statens offentliga utredningar* (1943: 39).
3. W. William-Olsson, *Ekonomisk-geografisk karta över Sverige* (Stockholm, 1946).
4. W. William-Olsson, *Economic Map of Europe* (Stockholm, 1953).
5. On request of the Norrland investigation the 1940 census of population tabulated data for all towns of 200 or more inhabitants. ("Utredning angående Norrlands näringsliv," p. 30). This made possible the more exact and detailed classification on William-Olsson's economic geographic map of Sweden published in 1946. The interest shown by Swedish geographers and city planners for the small agglomerations is a chief reason behind the publishing of data for such small areas by the 1945 and 1950 censuses.
6. Gerd Enequist, "Yrkesgruppernas fördelning i Sveriges kommuner år 1940," *Geographica* (1946).
Olov Jonasson, *Befolkningen och näringslivet i Mellansverige inom GDG:s trafikområde 1865–1940.* Minnesskrift utgiven av Trafikförvaltningen Göteborg-Dalarna-Gävle (Göteborg, 1949).
Yoshikatsu Ogasawara, *Japan, Labour Population and Urban Functions* (Map in the scale 1: 800000, published by the Geographical Survey Institute, Kurosunacho, Chiba, 1950).
H. J. Keuning, "Een Typologie van Nederlandse steden", *Tijdschrift voor Economische en Sociale Geografie* (1950).
L. L. Pownall, "The Functions of New Zealand Towns." *Annals of the Association of American Geographers* (1953).
Howard J. Nelson, "A Service Classification of American Cities," *Economic Geography* (1955).
J. F. Hart, "Functions and Occupational Structures of Cities of the American South," *Annals of the Association of American Geographers* (1955).

---

groups: (1) Cities and trade centers, (2) Manufacturing towns [(a) Mining and metal industry towns, (b) Forest industry towns, (c) Other manufacturing towns], (3) Railway towns. All towns of 200 or more inhabitants north of a line through Uppsala were classified according to this system and shown on a map with spheres in different colors. The system was further elaborated and presented on a map of all Swedish towns of 200 or more inhabitants.[3] After further modifications the classification system was applied on all European agglomerations of 10,000 or more inhabitants.[4]

Harris and William-Olsson approached the classification problem from somewhat different angles. Harris started out with some recognized types of cities and tried to find suitable criteria. His study is chiefly based on the population census and the censuses of manufacturing and business, but he also draws on other types of information. University towns, for example, are defined by a certain ratio of enrollment in schools of collegiate rank to the population of the city. William-Olsson, using only one statistical source, the 1935 census of population, made staple diagrams for individual towns showing the percentage of people employed in six main groups of industries. His classification was based on these diagrams.[5] On his map of Europe William-Olsson grouped the industries in two main categories, manufacturing and service industries. Among cities dominated by manufacturing nine different subtypes were distinguished.

Town classifications by Enequist (1946), Jonasson (1949), Ogasawara (1950), Keuning (1950), Pownall (1953), Nelson (1955) and Hart (1955) were, like William-Olsson's classifications, based on a uniform statistical soure, usually the population census.[6] Like Harris, these authors have seen manufacturing as one group of industries among others: trade, transportation, administration, etc. Ogasawara, who distinguishes the largest number of city types, adds an interesting element to the discussion by publishing frequency curves for his seven groups of industries. Nelson classifies the American cities with essentially the same technique as Ogasawara applies to the Japanese cities.

All the mentioned authors base their classifications on the total industrial structure without making any distinction between city forming and city serving ratios. They have to use different, more or less arbitrary, qualifications for their different types of cities. It is for example sufficient if a city has 4.4 per cent of its employment in Banking, insurance and real estate to be considered as a banking town by Nelson, but it should have 43.1 per cent in manufacturing to qualify as a manufacturing town.

In the present study "the city serving structure" was subtracted from the total structure before the classification was made. On the remaining "city forming structure" the same qualifications were applied to all industries: 5–10 per cent make a C-town, 10–20 per cent a B-town and more than 20 per cent an A-town (page 26).

## Statistical Sources

Several criteria can be used in quantitative studies of industrial distribution patterns. But if a comparison is essential only two measures will be considered seriously: employment and value added. The most commonly used yardstick and the one frequently accepted as the best single measure of economic activity is employment. This is the logical criterion if the object of the study is to analyse the population distribution. A series of maps based on employment, one for each of the industries, will directly show how people gain their livelihood. In this study cities are seen as popula-

tion agglomerations, as special phenomena in the general distribution pattern of population. This is in accordance with common usage. The importance of X-town is normally indicated by its number of inhabitants, not by its contribution to the national product or a similar criterion. Figures on employment have been used throughout this study, not as a substitute for something better, but as a direct measure of how people earn their living.

There are two sources of information on employment in many Western countries: the census of population and the censuses of different groups of industries. Only the first of these sources covers the total population. Large sectors such as public administration and different professional industries are not included in the latter source. There are also other differences. The census of population gets its information from individuals, the censuses of industries are based on reports from each establishment. The former source will have data classified according to the residence of the individual, the latter according to his working place. With the great majority of the urban population this does not make any difference since they work and live in the same city. If commuters do not make up a large share of those gainfully employed in a city and if they do not have an industrial structure which is significantly different from that of the total employed population the difference between classification according to residence or working place is unimportant in a structural analysis.

Both sources publish their data with breakdowns on industries and occupations.[1] In this study we are interested in the classification according to industries. We want to

1. *Industry* corresponds to the German *Erwerb,* the Swedish *näring* or *näringsgren,* the Danish *erhverv* and the Dutch *bedrijf. Occupation* is equivalent to *Beruf, yrke, fag* and *beroep.* Industry and occupation should not be used as synonyms.

know in what line of business people are employed, not what their occupations are. The census of population has a less detailed breakdown on industries than the other source, but it is detailed enough for our purpose.[1]

The censuses of industries have from our point of view another decisive defect, besides the incomplete coverage. A rule prohibits the disclosure of figures that would make it possible to calculate data concerning any one establishment. This rule makes the published employment data of the censuses of industries practically useless for quantitative geographic studies which go into any detail. The only information available for small administrative areas is the number of establishments. Maps prepared by the Bureau of the Census on different industries are based on these data, which do not give a fair quantitative impression of the industrial distribution patterns since plants vary considerably in size.

The present study is built on data contained in the *United States Census of Population, 1950, Series P-B*. The Census of Manufacturing for 1947 has been used as a complement for the verbal analysis of certain industries.

Our interest is focused on the urban agglomeration as an economic geographic unit, not on the corporate city as such. For cities having 50,000 inhabitants or more in 1940 or at a subsequent special census taken prior to 1950 the census gives data referring to three different city areas: the corporate city, the urbanized area and the Standard Metropolitan Area (SMA). For example Los Angeles, according to these three concepts, has 2.0, 4.0 and 4.4 million inhabitants respectively. The SMA, outside of New England, comprises one or more counties but never a fraction of a county. Where counties are large a considerable rural population as well as more or less independent towns may be included. The urbanized area consists of the contiguous, built-up surface, irrespective of corporate limits, and this most closely corresponds to the economic geographic city.[2] Wherever available, data on urbanized areas have therefore been used. The urbanized areas of the metropolises in general include a number of corporate cities, which thus are not considered in the present study as separate units.[3] In some instances data on urbanized areas were not given for cities with more than 50,000 inhabitants in 1950. In those cases data on the corporate city or the SMA were chosen.[4] For all cities with less than 50,000 inhabitants figures for the corporate city had to be accepted as representative of the geographic city. Small cities were grouped together if it was probable that they would have been considered as one urbanized area, had that concept been applied to cities with less than 50,000 inhabitants.[5]

1. For a detailed description of the coverage of each industry, see U. S. Bureau of the Census, *1950 Census of Population, Classified Index of Occupations and Industries* (Washington, 1950).
2. For detailed definitions of SMA and urbanized area, see U.S. Censuses of Population and Housing 1950, *Key to Published and Tabulated Data for Small Areas* (Washington, 1951).
3. Maps showing the extension of each urbanized area are available in the census publications.
4. The following cities are represented by their SMA: Bay City and Jackson, Michigan; Lima and Lorain-Elyria, Ohio; Muncie, Indiana; Lexington, Kentucky; Greenville, South Carolina; Gadsden, Alabama; Laredo, Lubbock, San Angelo and Wichita Falls, Texas; Ogden, Utah.
5. The following cities have been taken together: Auburn with Lewiston, Maine; Saco with Biddeford, Maine; Beacon with Newburgh, New York; Derby with Ansonia, Connecticut; Easton, Pennsylvania, with Phillipsburgh, New Jersey; Farrell with Sharon, Pennsylvania; Newsome Park-Hilton Park and Riverview with Newport News, Virginia; Bristol, Virginia, with Bristol, Tennessee; South Parkersburgh with Parkersburgh, West Virginia; Brownsville-Brent-Goulding and Warrington with Pensacola, Florida; West Monroe with Monroe, Louisiana; Texarkana, Texas, with Texarkana, Arkansas; West Lafayette with Lafayette, Indiana; St. Joseph with Benton Harbor, Michigan; Willow Run with Ypsilanti, Michigan; Springfield Place-Lakeview with

## Principles of Maps and Diagrams

*Figure* 1 shows all industries listed by the American census with bars, the widths of which are proportional to the employment. All bars are of equal length, representing 100 per cent. The placing of a bar with reference to the vertical zero line shows what percentage of the industry is rural and what percentage is urban. The diagram indicates, for instance, that about 94 per cent (94.2) of the employment in agriculture is rural and about 6 per cent (5.8) is urban. The rural population is divided into two categories: rural farm and rural nonfarm.

The data necessary for the construction of the diagram so far are directly available in the census publications.[1] The urban parts of the bars are divided into segments indicating different size groups of cities. The data for this part of the construction were obtained by adding the figures of the 864 individual cities, size group by size group, for the 36 urban industries.[2]

A key to the patterns is found at the bottom of the diagram. The key also shows how the total gainfully employed population is distributed between rural and urban and among different size groups of cities.

*Maps* 2–16 present the distribution patterns of fifteen manufacturing industries with circles for cities of 10,000 or more inhabitants having 150 or more people employed, and with uniform dots for employment outside of these cities. The circles are proportional to the employment. The dots represent 100 people employed in towns of less than 10,000 inhabitants and

in rural areas, as well as in cities with 10,000 or more inhabitants having less than 150 people employed. Two colors had to be used and the dots had to be given the heavier color to show through the circles where overlapping occurs. The dots were placed on a county basis. For some areas with large counties, especially some Rocky Mountain states, additional information was drawn upon to locate the dots as exactly as possible. The location of several dots in western United States, especially on the map of mining, should, however, be seen as approximate.

*Map* 1 shows all American cities of 10,000 or more inhabitants with squares which are proportionate to the size of the cities. The ten color gradings of the squares indicate the importance of manufacturing in individual cities according to the key in the cumulative distribution diagram.[3] This diagram shows, for instance, that ten per cent of the cities have a manufacturing percentage of 17.8 or lower, twenty per cent 22.0 or lower, etc. Ten per cent of the cities have a manufacturing percentage of 56.1 or

Battle Creek, Michigan; Marinette, Wisconsin, with Menominee, Michigan; Moorhead, Minnesota, with Fargo, North Dakota; Neenah with Menasha, Wisconsin; Urbana with Champaign, Illinois; Amphitheater with Tucson, Arizona; East Bakersfield, South Bakersfield and Oildale with Bakersfield, California; Seaside with Monterey, California; Alisal with Salinas, California; Pasco with Kennewick, Washington; Springfield with Eugene, Washington; Hoquiam with Aberdeen, Washington.

1. 1950 United States Census of Population. P–B 1 U.S. Summary. Table 55.
2. For the definition of cities used in this study, see page 23.
3. For the construction of the cumulative distribution diagram, see page 14.

---

FIG. 1. The United States had a gainfully employed population of 56,239,000 people in 1950. Their distribution on rural (rural nonfarm and rural farm) and urban (different size-groups of cities) is shown by the heavy line at the bottom of the diagram and this line also serves as a legend for the shadings of the bars. The sizes of the industries are indicated with figures at the left of the bars, the widths of which are proportional to the employment. The placing of a bar with reference to the vertical zero line shows what percentage of the industry is rural and what percentage is urban.

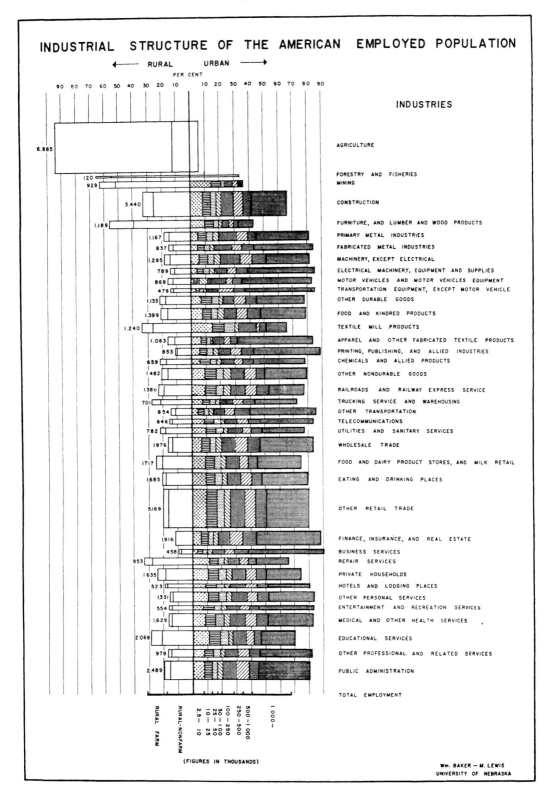

INDUSTRIAL  STRUCTURE  OF  THE  AMERICAN  EMPLOYED  POPULATION

← RURAL    URBAN →

PER CENT

90 80 70 60 50 40 30 20 10    10 20 30 40 50 60 70 80 90

INDUSTRIES

6.885    AGRICULTURE

120    FORESTRY  AND  FISHERIES
929    MINING

3.440    CONSTRUCTION

1.189    FURNITURE,  AND  LUMBER  AND  WOOD  PRODUCTS

1.167    PRIMARY  METAL  INDUSTRIES
837    FABRICATED  METAL  INDUSTRIES
1.295    MACHINERY,  EXCEPT  ELECTRICAL
789    ELECTRICAL  MACHINERY,  EQUIPMENT  AND  SUPPLIES
869    MOTOR  VEHICLES  AND  MOTOR  VEHICLES  EQUIPMENT
479    TRANSPORTATION  EQUIPMENT,  EXCEPT  MOTOR  VEHICLE
1.133    OTHER  DURABLE  GOODS

1.399    FOOD  AND  KINDRED  PRODUCTS

1.240    TEXTILE  MILL  PRODUCTS

1.063    APPAREL  AND  OTHER  FABRICATED  TEXTILE  PRODUCTS
853    PRINTING,  PUBLISHING,  AND  ALLIED  INDUSTRIES
659    CHEMICALS  AND  ALLIED  PRODUCTS

1.482    OTHER  NONDURABLE  GOODS

1.386    RAILROADS  AND  RAILWAY  EXPRESS  SERVICE
701    TRUCKING  SERVICE  AND  WAREHOUSING
854    OTHER  TRANSPORTATION
646    TELECOMMUNICATIONS
782    UTILITIES  AND  SANITARY  SERVICES

1976    WHOLESALE  TRADE

1717    FOOD  AND  DAIRY  PRODUCT  STORES,  AND  MILK  RETAIL

1,685    EATING  AND  DRINKING  PLACES

5,169    OTHER  RETAIL  TRADE

1,916    FINANCE,  INSURANCE,  AND  REAL  ESTATE
458    BUSINESS  SERVICES
953    REPAIR  SERVICES
1,635    PRIVATE  HOUSEHOLDS
523    HOTELS  AND  LODGING  PLACES
1,331    OTHER  PERSONAL  SERVICES
554    ENTERTAINMENT  AND  RECREATION  SERVICES
1,629    MEDICAL  AND  OTHER  HEALTH  SERVICES

2,068    EDUCATIONAL  SERVICES

978    OTHER  PROFESSIONAL  AND  RELATED  SERVICES

2,489    PUBLIC  ADMINISTRATION

TOTAL  EMPLOYMENT

RURAL  FARM

RURAL-NONFARM

2.5 — 10
10 — 25
25 — 50
50 — 100
100 — 250
250 — 500
500 — 1.000
1 000 —

(FIGURES  IN  THOUSANDS)

Wm. BAKER — M. LEWIS
UNIVERSITY  OF  NEBRASKA

higher, twenty per cent 49.9 or higher, etc. Since the two main groups, manufacturing and service industries, together make up 100 per cent of the gainfully employed population[1] the diagram can also be read the other way: ten per cent of the cities have a service percentage of 82.2 or higher, twenty per cent 78.0 or higher, etc.

*The relative maps* are drawn on the same base map as Map 1. Three types of towns are distinguished: A-towns, having a surplus of 20.0 per cent or more over the $k$-value;[2] B-towns, 10.0 to 19.9 per cent; and C-towns, 5.0 to 9.9 per cent. Cities having a surplus of 4.9 or less are not marked on these maps. The map of Other Retail Trade, showing all cities, is an exception.

1. The rural industries were excluded. Manufacturing includes mining and construction.
2. For a discussion of the $k$-value, see pages 17 ff.

# Manufacturing Industries

In the present study manufacturing includes mining and construction. Remaining activities are referred to as service industries. Manufacturing produces tangible goods, the service industries produce services. The distinction, which is also used by William-Olsson on his Economic Map of Europe, is significant from the geographic point of view. Most manufacturing industries are sporadic in their occurrence as a result of the haphazard distribution of mineral resources, the climatic influence on the differentiation of agricultural production and, perhaps most important, the advantages of specialization and large scale production. Although manufacturing employs a minority of the gainfully engaged American population, 33.7 per cent, it is much more important for an understanding of the uneven population distribution in the United States than the more or less ubiquitous service industries, employing no less than 52.4 per cent.

1. Sten De Geer, "The American Manufacturing Belt," *Geografiska Annaler* (1927).
Richard Hartshorne, "A New Map of the Manufacturing Belt of North America," *Economic Geography* (1936).
Helen Strong, "Regions of Manufacturing Intensity in the United States," *Annals of the Association of American Geographers* (1937).
Alfred J. Wright, "Manufacturing Districts of the United States," *Economic Geography* (1938).
Clarence F. Jones, "Areal Distribution of Manufacturing in the United States," *Economic Geography* (1938).
2. Fig. 1 suggests that it would have been justifiable to group Mining and the Furniture, and lumber and wood products industry with the rural industries. When this has not been done, it is because the arbitrary minimum limit set by the U.S. Census for

As is shown by Map 1 most cities with a high manufacturing percentage are located in the area north of the Ohio and east of the Mississippi or the Manufacturing Belt as outlined by Sten De Geer, Hartshorne, Strong, Wright and Jones.[1] In Iowa some manufacturing cities are located west of the Mississippi and in the Southeast there are two narrow extensions of the Manufacturing Belt, on the Appalachian Piedmont and in the Great Valley. In the Pacific Southwest Los Angeles stands out as an isolated large city with a high manufacturing percentage. On the Gulf Coast a small, outlying manufacturing region seems to be under formation.

In the decade 1940–50 cities with a high manufacturing percentage were on the whole growing slower than pronounced service cities (Table 2). The development was especially slow for cities located in coal mining districts. For further details, compare Map 1 and Map 17.

## Sporadic Manufacturing Industries

MINING

The American mining industry is the least urbanized of the urban industries. Only about one-third of the employees live in towns with more than 2,500 inhabitants. Over half of the mining population live in rural non-farm settlements,[2] chiefly small towns (Fig. 1). The industry can be

TABLE 2. Correlation Table: Manufacturing/Population Growth, 1940–50.
I–X Manufacturing, decils; A–J Population Growth, 1940–50, decils. For some reservations, see pages 83 ff. For calculation of decils, see page 14, and Maps 1 and 17.

|   | I | II | III | IV | V | VI | VII | VIII | IX | X | No. of cities |
|---|---|----|-----|----|---|----|-----|------|----|---|---------------|
| A | 2 | 4 | 5 | 5 | 11 | 8 | 12 | 10 | 12 | 16 | 85 |
| B | 5 | 4 | 6 | 5 | 7 | 10 | 7 | 12 | 12 | 17 | 85 |
| C | 5 | 6 | 6 | 6 | 7 | 7 | 13 | 12 | 12 | 11 | 85 |
| D | 1 | 4 | 2 | 9 | 6 | 9 | 12 | 14 | 14 | 14 | 85 |
| E | 6 | 6 | 2 | 8 | 15 | 2 | 15 | 13 | 7 | 11 | 85 |
| F | 6 | 10 | 6 | 10 | 6 | 12 | 7 | 13 | 9 | 6 | 85 |
| G | 11 | 9 | 12 | 12 | 11 | 9 | 6 | 5 | 7 | 3 | 85 |
| H | 11 | 15 | 19 | 13 | 10 | 10 | 1 | 1 | 3 | 2 | 85 |
| I | 20 | 18 | 9 | 9 | 9 | 11 | 3 | 2 | 3 | 1 | 85 |
| J | 18 | 9 | 18 | 8 | 3 | 7 | 9 | 3 | 6 | 4 | 85 |
| No. of cities | 85 | 85 | 85 | 85 | 85 | 85 | 85 | 85 | 85 | 85 | 850 |

divided into four branches: coal mining, with almost 50 per cent of the total employment, production of oil and natural gas (30 per cent), iron ore mining (4 per cent) and other mining (18 per cent).

## Coal

There is no strong correlation between coal reserves and actual output in the United States. The Appalachian field, extending from northwestern Pennsylvania to Alabama, dominates the American coal production. Outside of this district large-scale coal mining is carried on only in southern Illinois and adjacent states. The large coal deposits in the rest of the Interior Province and in the western regions only support a moderate mining industry.[1]

The important coal fields are all in, or fairly near, highly industrialized regions. The relations between coal-producing districts and industrial regions is of course not one-sided. Coal mining developed as a response to the demand for a cheap fuel, but manufacturing became concentrated in these regions because coal was available at low cost. The importance of the Appalachian

field for the development of the present distribution patterns of many American manufacturing industries can hardly be exaggerated. The importance of coal as a locational factor has progressively decreased with improved transportation and fuel technique and with the emergence of competing fuels, but it is still significant in certain industries, especially in the manufacturing of primary iron and steel.

Coal mining in the United States is carried on under favorable auspices, resulting in low costs. There is a definite connection between the rapid rise of the United States to the position as the largest manufacturing nation in the world and the steep increase of her coal output before World War I.

the classification of agglomerations as urban is very high, 2,500 inhabitants, and thus excludes a large number of smaller towns with definite urban characteristics. The corresponding minimum limit is as low as 100 to 200 inhabitants in several European countries. A considerable portion of the employment in bituminous coal mining, petroleum production and ore mining is found in towns with 500 to 2,500 inhabitants.

1. In a list based on coal reserves and expressed in bituminous coal equivalents four western states are placed at the top: Wyoming, North Dakota, Colorado and Montana. Erich W. Zimmermann, *World Resources and Industries* (New York, 1951), p. 464.

In the United States, where anthracite coal production has been and still is important, it is customary to make a distinction in the statistics between anthracite and bituminous coal.

Almost 100 per cent of the *anthracite* is produced in a small area in northeastern Pennsylvania with Wilkes-Barre and Scranton as the biggest centers. This area is favorably located, especially with respect to the densely populated region between Boston and Baltimore. Because of its smokelessness anthracite is a popular household fuel, but during the last decades it has met intense competition from fuel oil and natural gas. Because of the faulted and fractured seams (anthracite is a metamorphosed bituminous coal) the output per miner is considerably lower than in the bituminous industry. In other words, the anthracite industry has a far larger employment than its share of the total coal production would indicate. The output of an anthracite miner is roughly half as large as that of a miner in the bituminous branch of the industry. Wilkes-Barre and Scranton are the largest mining centers in the United States. There are also several smaller mining towns in this coal district.

1. The success of two German methods of making synthetic gasoline from coal, the Bergius process and the Fischer-Tropsch process, justify the hope that before long the outlook for the American coal industry might be brighter. As yet the manufacture of oil by these methods is more expensive than the world crude oil prices. Most American research in these fields is carried on by the petroleum industry, which recognizes the distinct supply advantages of coal over natural petroleum. The keen competition in the bituminous industry, with short-run survival as the necessary aim of most firms, has prevented accumulation of capital necessary for research which might provide dividends in the long run.
There has recently emerged a new dark cloud on the coal horizon: atomic energy. Available information indicates that this new source of heat energy means essentially a new fuel for steam power plants and thus in the next decades may prove to be a serious competitor of coal.
2. W. Adams, *The Structure of American Industry* (New York, 1950), pp. 37 ff.

Anthracite production exceeded the output of bituminous coal until about 1870. During a brief period preceding the establishment of Pittsburgh as the center of iron making, based on bituminous coal, a large portion of the American metallurgical industry was concentrated in the anthracite region. With a peak production of 100 million tons in 1917, the anthracite industry has since been on the decline. During the depression output dropped below 50 million tons and during World War II it never rose above 64 million tons. In 1950 production was 44 million tons. Employment has dropped from a peak of 180,000 in 1914 to 73,000 in 1950, with a resulting stagnation or decline in the mining communities. The anthracite region is a depressed area.

The American output of *bituminous coal* expanded very rapidly from the middle of last century: the average for the five-year period 1861–65 was less than 10 million tons, and the peak production in the First World War 579 million tons. This, however, seems to mark the turning point of the expansive stage of bituminous coal mining. From then on the long-term trend is downward, which is most clearly brought out by figures on the per-capita production.[1]

During the inter-war period the bituminous industry was sick. A large overcapacity, decreasing wages during the general prosperity of the Twenties, heavy losses for the industry as a whole even in the Twenties were some symptoms of the serious position of the bituminous coal industry, one of the most competitive in the American economic system.[2] A low of 310 million tons was reached in 1932, and not until the new war period were the production figures of World War I equaled and surpassed. The 1944 production was 618 million tons, a figure that was exceeded in 1947 (621 million tons), but in recent years the output has been considerably lower. In

1950 about 505 million tons were produced and in 1953 only 436 million tons. Employment in bituminous coal mining is decreasing rapidly.

The difficulties of the bituminous industry stem from the dwindling demand. Oil, natural gas and hydro-electricity have conquered a considerable share of the potential as well as of the conventional market for coal. The demand of the railroads has declined from 23 per cent of the national total in 1945 to 16.5 per cent in 1949, due to the displacement of coal-burning locomotives by Diesel engines.[1] The trend of coal for home heating seems to be definitely downward. Technical improvements in the coal-consuming industries, leading to a higher fuel efficiency, also have had a constricting effect.[2]

The bulk of American bituminous coal is mined in the Appalachian district, the biggest coal field in the world. It is estimated that normally over 90 per cent of the bituminous coal originates east of the Mississippi, and of this total, 70 per cent is produced in the Appalachian area, most of which lies in the dissected Allegheny Plateau, where natural conditions for coal mining are very favorable. Extraordinarily thick coal seams are found at relatively shallow depths, the beds are not deformed by faults and folding as in several European fields, and in the many river valleys the seams are often exposed, making possible the low-cost drift type of entries. The bulk of coal is taken from beds four feet thick or more. In recent years an increasing amount of coal has come from strip mines, in which the overburden has been stripped off by giant shovels. Strip mining accounts for about one-fourth of the total production. Where underground mining is used, the seams can often be reached by a tunnel (drift or slope), which is considerably cheaper than the vertical shafts commonly used in Europe, where the coal generally lies at great depths. Underground the coal is usually extracted by the room-and-pillar system, which is labor-saving but coal-wasting. About one-third of the coal is left in the form of pillars to support the roof. In Europe, where the longwall method is prevalent, only 5 to 10 per cent of the coal is left behind.[3] Because of the favorable geological conditions both cutting and loading in the mines are usually mechanized, whereas in Europe coal loading machines can seldom be used because of thin, distorted, steeply-pitching seams and poor roof conditions.[4] As a result of the complete mechanization in many mines, the output of coal per man-day is considerably higher in the American bituminous industry than in European mines. In the postwar years it has been over 6 tons. The United States still is a cheap-coal country in spite of the high wages which have in recent years been paid to the strongly unionized miners.

In the Appalachian field there has been a shift of the center of coal production from the northern states, Pennsylvania and Ohio, which dominated the markets until about the time of World War I, to the southern states, West Virginia, Kentucky and Virginia. There are several reasons for this shift to coal fields further away from the highly industrialized regions in the Manufacturing Belt. Coal could be produced cheaper in the South because of more favorable mining conditions than in the still undeveloped fields in the North and because of lower wages and lower land values. Since

1. E. L. Allen, *Economics of American Manufacturing* (New York, 1952), p. 241.
2. The average amount of coal burned to produce one kilowatt-hour of electrical energy in the United States in 1902 was 6.4 pounds, in 1920 3.0 pounds and in 1950 1.2 pounds.
3. Adams, *op. cit.*, p. 41.
4. William Van Royen and Oliver Bowles, *The Mineral Resources of the World* (New York, 1952), p. 18.

the end of the last century the middle and southern Appalachians have been connected with Great Lakes cities and with Norfolk by railroads specially designed to carry coal. Via Norfolk, the port cities of New England and New York City are supplied with cheap coal.

Only a small portion of the mining population in the Appalachian region lives in cities with more than 10,000 inhabitants, and the direct influence of mining on the industrial structure of these cities is smaller than might be expected. Coal mining has, however, attracted heavy industries with a large coal consumption to many towns. Several cities in the region function as service centers for areas that, thanks to coal mining, have a rather high density of population. Some of these cities rank in the highest decils for retail trade (see Appendix 1 and Fig. 29).

The outlying mining district in Alabama, centered on Birmingham, produces coal, iron ore and limestone for the large iron and steel industry of this area.

In the important mining district of southern Illinois and adjacent states, coal mining employs most of the miners, but oil production is also important. Illinois ranks among the ten leading oil states.

In the West coal mining is important in the Rock Springs district of Wyoming, in the Castlegate-Sunnyside area southeast of Salt Lake City, and in the Trinidad-Raton district on the border between Colorado and New Mexico. Utah, the largest coal producer west of the Mississippi, supplies the steel industry of California as well as its own industry. The Trinidad district sends coal to the steel center of Pueblo. All the western coal mines have had the transcontinental railroads as an important customer, but this is a shrinking market as the

number of diesel-electric locomotives increases.

## Oil and Natural Gas

Production of oil and natural gas in the United States started in the northern part of the Appalachian bituminous coal region. The first well was drilled near Titusville in northwestern Pennsylvania in 1859. Pennsylvania was for a long time the leading oil state and, later, adjacent Ohio and West Virginia joined as large oil producers. With the tremendously increased demand for petroleum products it became necessary to prospect for oil outside of this region, and very large fields were discovered in more remote parts of the country. About 1900 California entered the field as a leading producer, and about the same time midcontinental Texas and Oklahoma became high-ranking oil states.

The Appalachian district now produces only about one per cent of the national output. The decline has been both absolute and relative. The Midcontinent Province, stretching from eastern New Mexico, northern Texas and Louisiana through Oklahoma and Arkansas to Kansas, has for several decades been the leading district not only in the United States but in the entire world. Tulsa, Oklahoma, boasts of being "the oil capital of the world." The city has grown with the petroleum industry and serves, thanks to its central location, as a refining,[1] supply and administrative center for many oil fields. Numerous petroleum companies have their headquarters in the city and it is also a publishing center for oil and gas trade journals. Oklahoma City, focal point of a rich agricultural region, is located directly on an oilfield and derricks have been erected wherever space has permitted, even on the grounds of the state capitol. Shreveport is the center of an oilfield in northwestern Louisiana. Also Dallas, Texas, has directly

1. Oil refining is included in "Other nondurable industries."

and indirectly been stimulated by the oil industry. Several companies have their headquarters there. The relatively high density of cities in the Midcontinent Oil Province is intimately connected with the oil industry. Many of the smaller cities are mining towns of A-, B- or C-type (Fig. 2), and they have had the population base for their service industries substantially increased as can be seen from Map 2. Some of them have also attracted refineries (Map 14) and chemical industries (Map 13).

The second largest oil district, the Gulf Coast Province, includes parts of Texas, Louisiana and Mississippi with a small production also in Alabama and Florida. Almost all the oil comes from salt-dome traps. The fields are scattered on the coastal plain from the mouth of the Mississippi to the mouth of the Rio Grande. Oil was first found here in 1901 and the Gulf Coast Province soon became a major producer. During the last years this district has increased production very rapidly. With perfected scientific instruments, above all the seismograph, it has become possible to discover deeply buried salt domes. Prospecting meets with great difficulties in the coastal marshes, but with the help of conspicuous amphibious crafts, known as "marsh buggies," extensive explorations have been carried on. The oil deposits of the Gulf Coast do not stop at the shore line; large reserves are known off shore on the continental shelf. The ownership of the continental shelf (tidelands) has been a matter of dispute between the states and the federal government. Although the problem was solved in favor of the states through a law of 1953, there may be some hesitation on the part of the oil companies to make investments since the decision might be reversed with a change of administration.[1]

The Gulf Coast Oil Province is advantageously located with respect to ports. To these ports and to the large refineries in the port cities oil is also piped from the Midcontinent field. Crude oil and refined products are shipped from the ports (Baton Rouge, New Orleans, Port Arthur, Beaumont, Houston, Baytown, and Corpus Christi) chiefly to the Atlantic Seaboard but also to other parts of the world. Almost all the oil used in northeastern United States moves by tanker at a fraction of the cost of either tank-car or pipe-line transport.[2] Houston, one of the fastest growing big cities in the United States, is the business center of this expanding oil district.

California has for a long time been second only to Texas as an oil producing state. The main fields are located in the southern part of the state, in the Los Angeles Basin and in the San Joaquin Valley with Bakersfield as the chief center.

The Rocky Mountain Province, furthest removed of the oil districts from the principal markets, has large reserves, but production is still in the early development stage. Wyoming, Colorado, and Montana supply the northern Great Plains and the intermontane region with oil. Casper, Wyoming, is an oil town.

Natural gas is found in most oil fields. Much still goes to waste and a large portion is used close to the wells in oil field operations or in industries such as carbon black production and petroleum refining. Through an extensive pipe line system natural gas has, however, been made available to a large part of the American population. It is now a major industrial and domestic fuel even in areas far removed from the oil fields, and production of natural gas is expanding faster than that of the other main fuels. With the emergence of natural gas as a

1. C. L. White and E. J. Foscue, *Regional Geography of Anglo-America* (New York, 1954), pp. 198 ff.
2. W. E. Pratt and D. Good, *World Geography of Petroleum* (Princeton, 1950), p. 146.

# Mining Towns

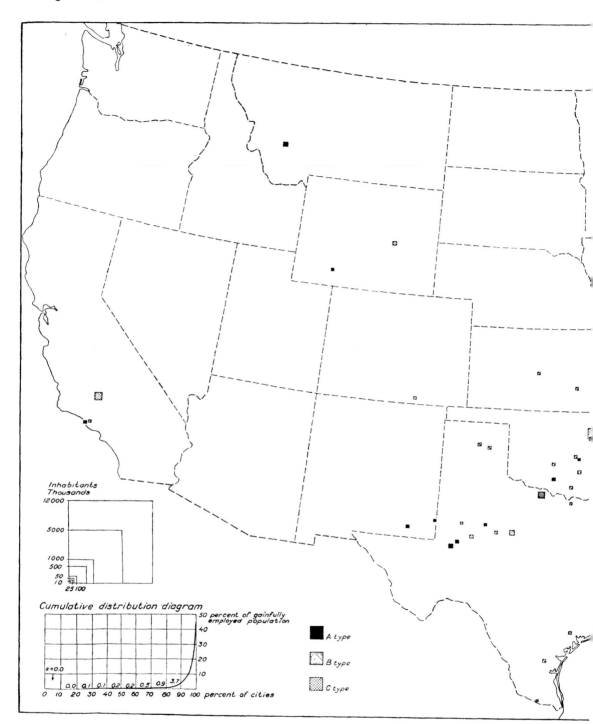

FIG. 2. The largest concentration of mining towns is found in the small anthracite region of northeastern Pennsylvania with Wilkes-Barre and Scranton as leading centers. Bituminous coal mining, much more important than anthracite mining, is of little consequence as a city forming activity. Oil and natural gas mining supports a large number of small mining cities in Texas, Oklahoma, Kansas and other states.

major American fuel, fields have been opened which contain mainly gas (dry fields).

## Iron Ore

Iron ore mining in the United States is dominated by the Lake Superior region, which normally produces four-fifths of the total output.[1] In second place is the Birmingham district with 7 to 8 per cent.

The iron ore fields in Minnesota and Michigan, with their relatively small number of employees, have not attracted any considerable urban settlement. The biggest mining centers of the Lake Superior region, Hibbing, Virginia and Ironwood, only have between 11 and 16 thousand inhabitants. This is in sharp contrast to the coal districts, which in the United States as well as in Great Britain, Germany and other countries have attracted a tremendous urban population. Iron-ore fields exert a pull only on the primary iron and steel industry, but coal deposits also attract other types of heavy manufacturing, among which are important steel-consuming industries. This leads to an agglomeration of urban industries on or near the coal fields, which thus get the added attraction of a big market, ranking with coal and iron ore as the most important locational factor in the steel industry.

The largest production of iron ore in the West occurs in the Iron Mountain area of southwestern Utah and in the Eagle Mountain district of southern California. Utah ranks fourth in iron ore output in the nation with a production of about four million tons, half of which is consumed in the state. The rest is shipped to Pueblo, Colorado, and Fontana, California. The output of the Eagle Mountain district goes to Fontana.

## Other Mining

Copper mining employs more people than any other nonferrous ore production. The well-known copper town, Butte, Montana, is bigger than any iron-ore city in the United States. The gigantic Bingham Canyon mine, just outside of Salt Lake City, is the largest copper producer in the country. In Arizona, the leading copper state, there are several big mines: Bisbee in the southeast on the Mexican border, Ajo further west and also near the national border, Globe-Miami in the center of the state with Mineral Creek in the south and Jerome in the northwest, and finally Morenci near the border of New Mexico.[2] The miners in these copper districts live in towns of less than 10,000 inhabitants. Other large copper districts are Santa Rita in southwestern New Mexico, Ely in eastern Nevada and the Upper Peninsula in Michigan. The latter field was the leading American copper district for a long time in the last century.

A small area in central Arkansas accounts for almost the whole production of bauxite in the United States. Lead and zinc mines are widely scattered over the continent. The most important are the Tri-State area in Oklahoma, Kansas and Missouri, the Coeur d'Alene district in northern Idaho and the Leadville region in the Colorado Rocky Mountains. Primarily, lead is mined at Bonne Terre in Missouri, south of St. Louis, and zinc at Franklin Furnace in New Jersey, east of New York. Some of the lead and zinc are by-products of the copper mines.

Gold is produced in the Homestake mine in the Black Hills of South Dakota, the largest gold mine in the Western Hemisphere, in the Cripple Creek district of the Colorado Rocky Mountains, and in a wide area on the western slope and in the foothills

1. Iron ore production in the United States reached a peak in 1917 with 75 million tons, an output that was not equaled until World War II. The output is very sensitive to business fluctuations, as indicated by the 1932 production of 10 million tons. In the war and postwar period production has oscillated between 70 and 105 million tons.
2. Van Royen and Bowles, *op. cit.*, p. 100.

of the Sierra Nevada in California, the gold field of the 1848 Gold Rush. Silver is associated with other metals—copper, lead and gold—and most of the output comes from fields that have already been mentioned.

Four-fifths of the American phosphate rock production originates in western Florida, most of it near Tampa. The Carlsbad district of New Mexico dominates the potash output. Large deposits of high-grade phosphate rock and potash are found in scattered localities in the West, but because of the long distance to the large fertilizer markets in the Middle West, South and Middle Atlantic States, production is small.

The sulphur districts do not stand out on Map 2, since almost the whole output is produced in coastal Texas and Louisiana, where the sulphur is extracted by the Frasch process from cap rocks overlying certain salt domes. The sulphur district thus merges with the Gulf Coast Oil Province. Most of the sulphur-producing salt domes are found just west of Houston-Galveston, and Galveston is the leading shipping port for sulphur.

FURNITURE, AND LUMBER AND WOOD PRODUCTS

The industries using wood as their raw material are represented on Map 3 with one exception: the wood pulp industry, which is included on the map for "Other nondurable goods industry." Sawmills, including logging, and related factories like veneer mills, plywood plants and box factories, employ about two-thirds of the people in this industry—here referred to as wood manufacturing—and the furniture industry the remaining one-third.

Wood manufacturing, to a very high degree, is located in small towns and in rural districts as can be seen from Fig. 1 and from the many dots on Map 3.

The general distribution of the industry closely follows the distribution of forests. This coincidence is evident even if the comparison is made with the virgin forest pattern[1] of the earliest period of white settlement on the North American continent.[2] Wood manufacturing, thus, is raw material oriented. The dots are, however, unevenly distributed. They are denser in two areas: in the South from Virginia to eastern Texas, and in the northern part of the Pacific Coast region.

A series of wood manufacturing maps showing in time sequence the distribution of the industry throughout the pioneer period up to the present would of course show a movement west and south, but it would also indicate areas where the industry has thinned out after a brief period of frenzied production, either because the soil has been cleared for agriculture or because of wasteful cutting practices in areas with a climate too harsh for farming, where forestry on a sustained yield basis might have supported an important lumber industry and thus a denser population than actually lives there. The Great Lakes region can be used to illustrate both these points. Many cities in Michigan, Wisconsin and Minnesota were for a brief period during the last century known as lumber towns. Pioneer farmers in these states got a substantial part of their cash income from the sale of forest products when the expanding railroad net had connected their county

1. A map published by the Forest Service of the U.S. Department of Agriculture and reproduced in American textbooks in social sciences, shows the areas covered by virgin forest in 1620. See for instance Zimmermann, *World Resources and Industries* (New York, 1951) p. 403.
2. The distribution of tree growth was primarily influenced by precipitation. Forests in the United States are limited to areas of at least 20 inches of annual rainfall and no long periods of drought. The differentiation of forest forming species within forested land of similar climate seems to be determined mainly by soil conditions. J. R. Whitaker and E. A. Ackerman, *American Resources* (New York, 1952).

# MINING

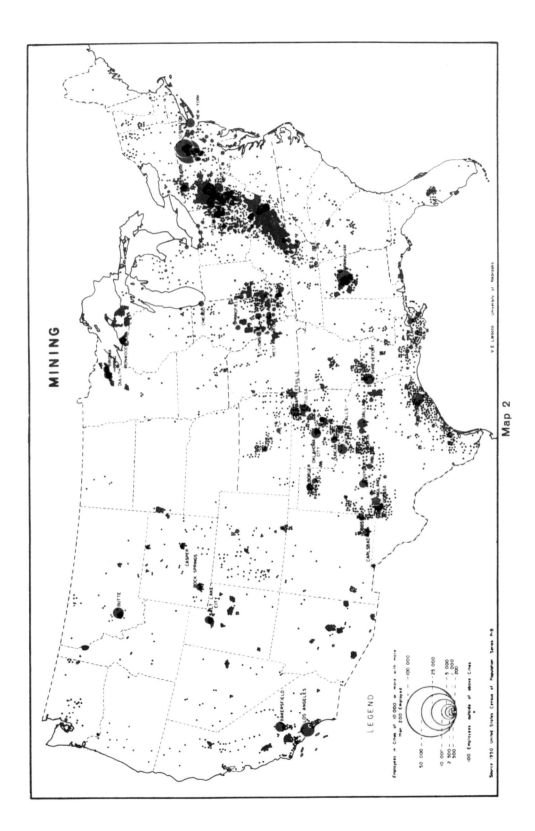

Source 1950 United States Census of Population Series P-8

Map 2

V. E. Lawson    University of Nebraska

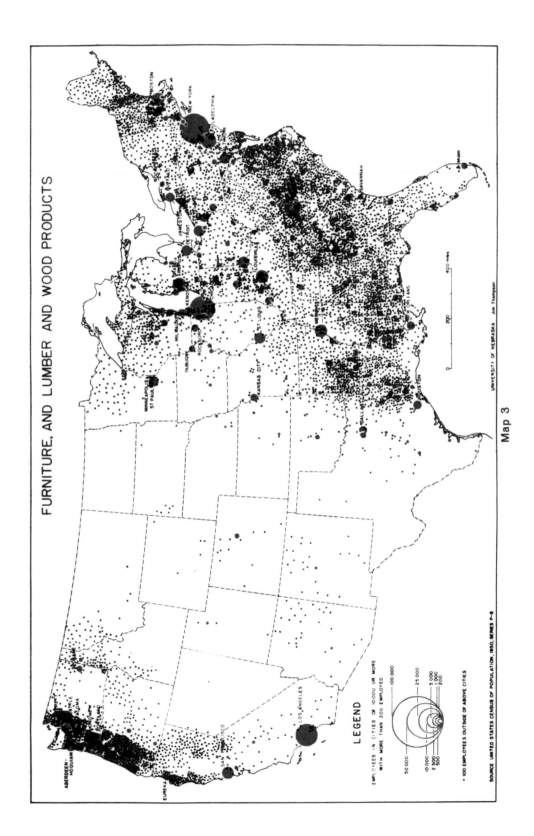

FURNITURE, AND LUMBER AND WOOD PRODUCTS

LEGEND

EMPLOYEES IN CITIES OF 10,000 OR MORE
WITH MORE THAN 200 EMPLOYED

100,000

50,000

25,000

10,000
5,000
2,500
1,000
500
200

• 100 EMPLOYEES OUTSIDE OF ABOVE CITIES

SOURCE: UNITED STATES CENSUS OF POPULATION, 1950, SERIES P-8

UNIVERSITY OF NEBRASKA    Jim Thompson

0      200      400 miles

Map 3

with the ever-increasing markets. For nearly a century lumbering dominated all other industries in the Upper Lakes region. But the "mining" of the forest left in its wake ghost towns and stripped land—a poor basis for a longtime economy, especially where soil and climate are not suitable for farming. Because of these shortsighted exploitation methods which have been characteristic of American forestry almost everywhere, the present distribution pattern for wood manufacturing shows an anomalous situation. The northeastern, originally forested part of the country (the Manufacturing Belt and adjacent areas) with its dense population and big market, has a considerably lower wood manufacturing density than such remote forest areas as the Southeast and the Pacific Northwest.

## Lumber and Wood Products

The lumber industry is dominated by small plants. With an output of less than 5 million board feet each, 38,000 small mills account for nearly half of the total production. Most of them are primitive portable mills. There are only a little more than 200 large mills with a production of more than 25 million board feet each, but they produce one-third of the total output.[1]

The Pacific Coast lumber industry is almost entirely a producer of softwoods. Despite a successive diversification in the economy of this area, the lumber industry is still the leading branch of manufacturing in Washington, Oregon and northern California. The natural conditions for lumbering are the best possible in this region. Precipitation is high since the coastal ranges are in the path of the cyclones during most of the year. The amplitude of yearly temperature extremes is moderate or low. There is a pronounced drought period during the summer, when the cyclones follow a more northerly path, but it is not so long as it is further south. This dry period makes forest fires a big problem. Serious as it is for the long-run economy of the area, its importance for the short-run profit interest should not be exaggerated. The burnt-over land does not represent a total loss as the trunks are left rather intact and tops and branches might not have been utilized anyway.

The timber volume per acre is probably higher than in any other region in the world: the trees are big, the stands are dense and virgin timber still predominates. In comparison with forests in northern Sweden the Douglas fir areas of the Pacific Coast contain about four times as much timber per acre.[2] On 13 per cent of the United States' commercial timber land, the Pacific Coast district has over half the national saw timber volume.

Most of the intense lumber district lies within the Douglas fir region and this tree dominates the total cut. In the coast region of northern California the gigantic redwoods support an intense saw mill industry, but in the Pacific Coast region as a whole Ponderosa pine ranks second to Douglas fir.

Although the saw mills of the Pacific Coast are far away from the large markets, many are readily accessible to the ocean and cheap water transportation, especially in the Puget Sound area, along the lower course of the Columbia River, at Coos Bay and around Grays Harbor. Most of the big plants have a tidewater location.[3] The rug-

1. E. W. Zimmermann, *World Resources and Industries* (New York, 1951), p. 416.
2. E. Rostlund, "Nya utvecklingslinjer och gamla ovanor i väst-amerikansk skogsindustri," *Ymer* (1954), p. 44.
3. Exports used to be an important factor in the Douglas fir region, but during the past 20 years lumber for domestic construction has determined the level of activity in the Pacific Coast lumber industry. The Midwestern market is very important which partly explains that in 1949 over 80 per cent of the Douglas fir sold in the USA were shipped by rail. "The Lumber Industry of the Pacific Coast." Supplement to *Monthly Review* (December, 1950). Federal Reserve Bank of San Francisco. P. 19.

ged terrain, as well as the size of the logs, offers difficulties which have been overcome with suitable techniques, involving more mechanization than in the rest of the country. It is therefore natural that the average size of saw mills should be greater than in the country as a whole. This does not mean, however, that the lumber industry is confined to big saw mills. Even if there are examples of big integrated plants, combining saw mills and planing mills with pulp mills and other factories which can utilize waste, the characteristic saw mill even of this region is the medium-sized and small plant with its huge, conic burner for disposing of waste.

In the Pacific Coast region there are among cities with more than ten thousand inhabitants several examples of one-sided saw-mill towns, and most of the other cities are strongly influenced by lumbering. This is true also of the big cities, Seattle, Tacoma and Portland. Their industrial structure is diversified, but wood manufacturing in all three is a major manufacturing industry.

A good example of a one-sided sawmilling town is Longview, Washington, on the lower Columbia River. What is now Longview was chosen as a suitable location for a big, integrated plant in 1922, and two years later there was a city of five thousand inhabitants. Longview has grown up around this establishment, which now consists of three sawmills, a planing mill, a sulphite pulp mill, a sulphate pulp mill utilizing waste and low grade logs, a plywood plant, a plant for compressing sawmill refuse into small logs for fire places, a bark conversion plant, and a central power plant. In 1950 Longview had 20,000 inhabitants. It is one of the best planned cities on the American continent. Other one-sided lumber towns are the twin cities Aberdeen and Hoquiam (Grays Harbor), Washington, Bend, Oregon, and Eureka in the California redwood

district. None of the towns in the important lumber center Coos Bay, Oregon, reaches the size of 10,000 inhabitants.[1]

The southeastern lumber district stretches from Virginia into eastern Texas. Large-scale exploitation started here at about the same time as on the Pacific Coast: around the turn of the century, when the forests of the Upper Lake region had been almost depleted. The timber volume per acre in the virgin stand was here considerably lower than on the Pacific Coast because of smaller species, but soil and climate are very favorable for a rapid growth. Forest exploitation in the South has followed the general pattern of migratory lumbering, the so called "cut out and get out" sequence. As a result very little virgin forest remains and the lumber men have left the forest in a poor condition. Lumbering has been favored by the topography, which in the coastal plain is very level, and by cheap labor. Here, as in the other forest regions, lumbering is slowly changing from a migratory to a permanent industry with forestry conducted on a sustained-yield basis.

Production of softwoods dominates, but hardwoods are also important. In a broad belt along the Atlantic and Gulf Coast, four species of yellow pine are the forest forming trees on the prevailing sandy soils with hardwoods and cypress in the river bottoms. The largest area of hardwoods, one of the most important in the world, is found along the lower Mississippi River. The Appalachian Piedmont region, inland of the coastal sand plain, and the Ozarks are characterized by hardwoods, chiefly oak, but with intermingled stands of yellow pine.

In the South there are no lumber towns with more than 10,000 inhabitants, but lumbering is the most important manufacturing industry in several of the Southern

1. O. W. Freeman and H. H. Martin, *The Pacific Northwest* (New York, 1954).

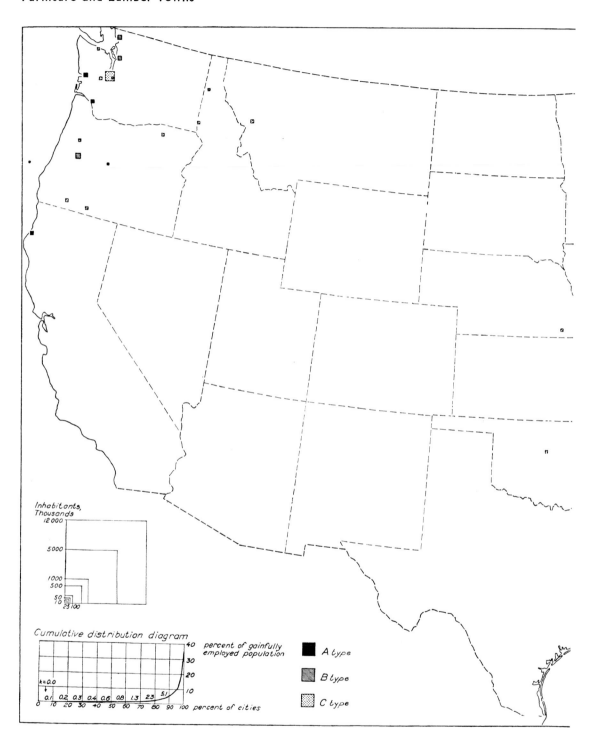

FIG. 3. Lumber and furniture towns are as a rule rather small and they are scattered over the continent. Grand Rapids, Michigan; Jamestown, New York; Gardner, Massachusetts; and High Point, North Carolina can be mentioned among the furniture towns, Longview, and Aberdeen-Hoquiam, Washington; Springfield, Oregon; and Eureka, California, among the lumber towns.

cities, especially in Mississippi. The conspicuous concentration of wood manufacturing in the Piedmont region of North Carolina and southern Virginia is due to a large furniture industry.

*Furniture*

The furniture industry in the United States, as in other countries, is composed largely of small units. Big factories are not markedly more efficient than plants of moderate size.[1] The industry turns out a bulky and relatively cheap product which in general cannot very well carry the costs of long-distance transportation. As a result of these two characteristics, furniture manufacturing is widely spread over the continent but the factories tend to cluster in furniture towns and there is a certain specialization among them: in one town mostly bedroom furniture are made, in another cabinets and in a third chairs, etc.

Chicago, New York and, more recently, Los Angeles are important furniture centers, indicating a market orientation but also an orientation to the fashion centers.[2] The furniture industry of big cities differs in character from that of small towns. Up-

1. The predominance of small plants is reflected in the design of the products. Whereas the automobile industry, early combined in a few big companies, soon discarded the buggy as a prototype and made radical new designs, the industrialization of furniture making did not have this effect. The small American furniture manufacturer did not and could not make such a break with tradition and adapt his designs to new materials and new production methods. He could not afford to hire skilled design personnel or establish his product through consumer advertising. As a result the industry looks back through the centuries for its production prototypes. Only the home building industry—also made up of small units—can be compared with furniture manufacturing in this respect.

G. Nelson, "The Furniture Industry", *Fortune* (1947).

2. The style-forming influence of the movie industry must be great. It can hardly be a mere coincidence that Los Angeles during the last few decades has become an important furniture center, an apparel center and a style center for modern American architecture (California bungalows, "ranch" houses).

holstered products make up a disproportionately large share. New York City is by far the largest producer of upholstered furniture in the country. Much of the wooden furniture work in the big cities—employing more people than upholstering—is custom work made to special order and not distributed outside the metropolitan areas.

The oldest and best known American furniture city is Grand Rapids, Michigan. This city was an important lumber town in the last century and the furniture industry was an outgrowth of lumbering. At the end of the nineteenth century Grand Rapids had the largest production of factory-made furniture in the United States and probably in the world. As furniture quality in the beginning was incompatible with factory production, the term "Grand Rapids" has come to mean furniture of bad taste and quality. When the local lumber supplies ran out around 1920, Grand Rapids began to lose its position but it still retains its leadership in furniture styling. In the trade Grand Rapids now is best known for its high-quality furniture. The city is a diversified manufacturing town but furniture still is the leading city forming industry.

The rapidly expanding wooden furniture industry of the South originally was aimed at the local market. The major producing area is located in the Piedmont section of North Carolina and southern Virginia with High Point as the best known center. On Map 3 this furniture district is conspicuous by a remarkable concentration of dots and circles. Many of the small shops and factories are located in towns smaller than 10,000 inhabitants. The manufacture of furniture began here in the 1880's during the Reconstruction of the South. Cheap factory-made household furniture and the coarser cotton fabrics had a substantial regional market. The raw materials were also available as well as a huge supply of cheap labor. It was natural for the Southern industrialists to direct their interest toward this type of product, cheap necessities which could be fabricated even by unskilled workers and therefore are among the first items taken up by agricultural regions and nations with manufacturing ambitions.

With some decades of experience, it became possible from the early 1920's to capture markets outside of the South.[1] The furniture district of North Carolina and Virginia now includes some of the largest and most advanced plants in the country. There are examples of vertically integrated companies which own forest tracts, exploited on a sustained-yield basis, and sawmills, in addition to the furniture plants. The industry has, however, outgrown its local timber basis and much lumber is shipped in from other states.

The only wood manufacturing city of A-type in this region is Thomasville, with one of the largest chair plants in the world. Several of the other cities are strongly influenced by furniture manufacturing: High Point, Lexington, Goldsboro, Statesville and Hickory in North Carolina and Martinsville in Virginia. The leading position of High Point within the Southern furniture district is marked by a huge exhibition building. Permanent exhibitions are held here, just as in New York, Chicago, Los Angeles, Grand Rapids and Jamestown. In these cities are also held two annual markets, when the manufacturer shows his products to the retail market.[2]

Jamestown, New York, one of the oldest furniture towns in the United States, specializes in case goods of wooden household furniture and in office furniture. Gardner, Massachusetts, one of the few cities dominated by furniture manufacturing, is the home town of one of the largest firms in the industry. Two Rivers, Wisconsin, another furniture town of A-type,

specializes in dental cabinets and printer's furniture. Louisville, Kentucky, became an important furniture center when the city was chosen as the location of one of the largest furniture firms in the United States. In Indiana, one of the leading furniture states, there are no conspicuous clusters of plants.

PRIMARY METAL

The distribution patterns of the various manufacturing industries in the United States are very different. If they all were synthesized on a map of the manufacturing industries in general, the Manufacturing Belt in the northeastern part of the country would stand out. Only a few industries have, however, a distribution pattern which largely coincides with, or almost entirely falls within, the Manufacturing Belt. They are all, with perhaps one exception,[3] branches of metal manufacturing. If a map were made for the latter only, the Manufacturing Belt would stand out with very distinct limits, much clearer than on a general manufacturing map on which the non-metallic industries with their diverging distributions blur the pattern. The metal manufacturing industries are the substance of the Manufacturing Belt, especially in its western part, west of the Alleghenies.

The location of the primary metal industry[4] has been of great importance for the general location of the different branches of metal manufacturing, which turn the metals,

---

1. B. F. Lemert, "Furniture Industry of the Southern Appalachian Piedmont," *Economic Geography* (1934).
2. Best known of the exhibition buildings is the American Furniture Mart in Chicago, a sixteen-floor building with a hall which is an exact copy of an English cathedral.
3. "Other durable goods industries" also have a similar distribution pattern.
4. Blast furnaces and steel mills, iron and steel foundries, nonferrous metal production, including rolling, drawing, and foundries, wire drawing, etc.

mainly iron and steel,[1] into finished products. This must be true even if there evidently exists a high degree of interdependence between primary and secondary metal manufacturing. The location of the secondary metal manufacturing branches in close contact with the primary metal industry offers advantages in transportation costs, which are an important cost item for the heavy and relatively cheap products which make up the bulk of the output of the primary metal industry. The metal working industries together constitute a considerable portion of the manufacturing industries of the United States (Fig. 1), and it is thus evident that the primary metal industry holds a key position to an understanding of the general population distribution of the country.

Map 4 suggests the following grouping of the American primary metal centers:

1. The biggest concentration in the United States and at the same time on the American continent stretches like a broad, straight band in a northeasterly direction from southwestern Pennsylvania (Pittsburgh, Johnstown, and several small cities) through northern West Virginia (Wheeling, Weirton) northeastern Ohio (Steubenville, Canton, Youngstown, Cleveland, Lorain-Elyria) to eastern Michigan (Detroit, Saginaw). An outlier of this district extends in an easterly direction along the shores of Lake Erie (Buffalo, Niagara Falls, Erie). The outlined area accounts for about half of the American steel capacity.[2]

2. The second most important district is located on the southern shores of Lake Michigan with Chicago as the dominant center (includes the Calumet district and Gary). About 20 per cent of the American steel is produced here. — St. Louis has a much smaller but still considerable production of primary metal products.

3. The Atlantic Coast from Baltimore to New York, including eastern Pennsylvania. This area extends into New England. It accounts for less than 20 per cent of the steel produced in the United States. Of the big cities Allentown-Bethlehem and Waterbury are most influenced by the primary metal industry.

4. The Southern district, chiefly Birmingham, Gadsden and Anniston in Alabama, with about 5 per cent of the national steel capacity.

5. Scattered centers in the West, which total about 5 per cent of the American steel output. Los Angeles is the most important production center.

Pittsburgh has for a long time—since the Civil War—been the leading steel center on the American continent. The proximity and easy access through river transportation to great resources of high-grade coal, especially the unexcelled Connelsville coking coal, has been the decisive factor in giving Pittsburgh this prominent position within the world's biggest steel industry. The city is located at the confluence of the Monongahela and the Allegheny Rivers, where they form the Ohio River. The deeply entrenched, rather small Monongahela River, studded with steel plants, is one of the world's busiest water routes, averaging more than 30 million tons per annum. In the early years the Pittsburgh steel industry used local iron ores but from the 1880's Lake Superior ore was made available through low-cost lake-rail transportation. Coal, coke and finished steel products moving north furnish return freight for the ore cars moving south.

The policy of the gigantic US Steel Cor-

1. On iron and steel falls more than 80 per cent of the total employment in the primary metal industry. The nonferrous metals (copper, aluminium, zinc and lead) are of secondary importance.

2. Edward L. Allen, *Economics of American Manufacturing*. (New York, 1952), p. 85.

poration[1] helped Pittsburgh retain and develop its prominence. With the elimination of the "Pittsburgh Plus" in 1924 and of the multiple basing-point system in 1948,[2] competing steel centers were in a better position and it has long been evident that the expansive steel industry of Chicago would ultimately compete with Pittsburgh for the position as leading steel center. In absolute figures the primary metal industry now employs about the same amount of people in Pittsburgh (106,000) and Chicago (114,000), but as the former city has only about one-third as many inhabitants as the latter it is evident that Pittsburgh's economy is dominated by steel to a much higher degree (18.5 per cent; B-type) than that of Chicago (5.4 per cent; C-type).

Chicago has none of the raw materials for steel production in its vicinity. Its greatest asset as a steel center is a favorable transportation situation with regard to raw materials and markets. Coal is drawn from the Appalachian field and from southern Illinois, ore from the Lake Superior fields. The steady westward push of the center of population and with it the center of metal manufacturing industries has worked into the hands of Chicago recently, just as it did with Pittsburgh in an earlier period. Chicago itself is prominent in all of these industries.

The other primary metal centers in the dominant district south of Lake Erie have a position similar to that of Pittsburgh or Chicago. The inland cities have, like Pittsburgh, low transport costs for excellent coking coal of the Appalachian field but have to pay more for lake-rail-transported ore from the Lake Superior than the shore cities. These in their turn have higher transport costs for coal than the inland cities.[3] Like Chicago the shore cities have an excellent market location, as can be seen from the maps of different metal manufacturing industries. Somewhat puzzling is,

however, the "underdeveloped" steel industry of Detroit. Automotive manufacturing takes a considerable portion[4] of the total American steel output and it would motivate a bigger concentration of steel manufacturing to Detroit, the general location of which must be as favorable as that of Chicago. Detroit is, however, rapidly increasing in importance as a steel center.

An interesting outlier of the Great Lakes steel industry is the plant in Duluth, shipping port for the Mesabi Range ores. A plant was built in this remote, sparsely populated region by the US Steel Corporation in 1915 after pressure from the people of ore-exporting Minnesota.[5] Similar considerations were taken when new steel plants were built by the Swedish Government at Luleå (northern Sweden) and by the

---

1. US Steel has about $^1/_3$ of the American ingot capacity. E. L. Allen, *op. cit.*, p. 83.

2. Under the "Pittsburgh Plus" system all steel companies regardless of location quoted on a given order the price set by the US Steel Corporation at Pittsburgh and added to this the transportation cost to the consumer, even though the steel might be made at some other point. In 1924 the Federal Trade Commission ordered the producers to abandon the system and as a result there was a gradual increase in the number of basing points. In 1948 the steel industry abandoned the basing point system altogether and shifted to f.o.b. prices after the Supreme Court had ruled against the use of a basing-point system in the cement industry.

3. For estimated assembly costs in the production of pig iron at selected points, see Alderfer and Michl, *Economics of American Industry* (New York, 1950), p. 68.

4. Detroit industries consume more than 15 per cent of the American steel production, but only 5 per cent of the steel capacity is located in the region. Automotive manufacturing normally is the largest steel consuming industry taking 15–20 per cent of the finished steel production. "Construction and maintenance" also takes about 15 per cent of the steel. During some war years shipbuilding was the largest steel consumer, but it is normally rather unimportant with about one per cent of the steel consumption. E. L. Allen, *op. cit.*, p. 77.

5. Minnesota had a good negotiation basis in its right to levy taxes on the ore. The steel company's hesitation seems to have been justified if we may judge by the slow growth of this steel district. L. White and G. Primmer, "The Iron and Steel Industry of Duluth: A Study in Locational Maladjustment," *Geographical Review* (1937).

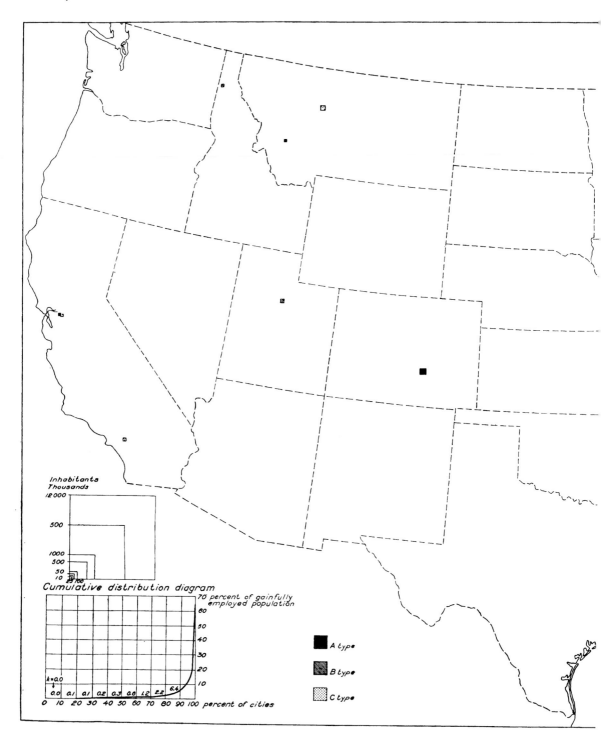

FIG. 4. Several of the largest American cities are primary metal towns: Chicago, Pittsburgh, Cleveland, Baltimore and Buffalo.

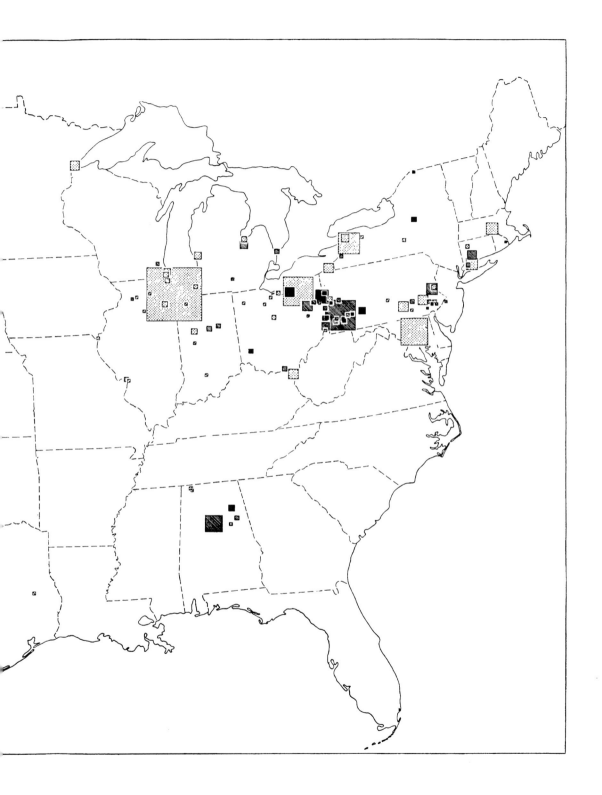

Norwegian Government at Mo in Rana (northern Norway). These locations are questionable in spite of low assembly costs for raw materials at or near the ore shipping ports with their favorable return freights for coal and coke and with cheap hydro-electricity available, because the finished products have to be shipped long distances.[1]

The Atlantic Coast district offers the steel industry, beside its big market, the advantages of cheap imported ore and cheap market scrap,[2] which accumulates in this highly industrialized area. It is the oldest steel district in the country. During a brief period from 1840, when the industry turned from charcoal to anthracite, the valleys of eastern Pennsylvania came into their own. With the transition to bituminous coking coal and Lake ores came a relative decline for the whole district. A large share of the production of this area is in the hands of the Bethlehem Steel Corporation,[3] with principal plants in Baltimore (Sparrows Point) and Bethlehem. This company also has large plants in Buffalo and Johnstown.

1. It will take several more years before the success of the two Scandinavian mills can be judged. In the meantime it is wise to see them as national efforts to create employment in peripheric and economically underdeveloped but strategically important regions, known as communist strongholds. If they meet with economic success, a revaluation must be made among economic geographers about the importance of different locational factors in the steel industry.

2. About as much scrap as pig iron is used by the steel industry. Of 55 million tons used in 1950, close to 25 million tons were "purchased" scrap and the remainder originated in the steel industry (home scrap). *The Making of Steel*, American Iron and Steel Institute (New York, 1951), p. 22.

3. Bethlehem Steel has about 15 per cent of the American ingot capacity. E. L. Allen, *op. cit.*, p. 83.

4. Among them US Steel Corporation's Fairless Plant at Morrisville, Pennsylvania, on the Delaware River. It is based on imported ore (chiefly from Venezuela) and has an ingot capacity greater than that of Japan or Sweden (1.8 million tons). *Industrial Cities Excursion, Guidebook*. XVIIth International Geographical Congress (Washington, 1952), p. 37.

5. George H. T. Kimble, "The Geography of Steel," *Scientific American* (1952), p. 52.

6. White and Foscue, *Regional Geography of Anglo-America* (New York, 1943), p. 422.

During the last decade the two biggest American steel companies have been actively engaged in successful ore prospecting in various parts of the world as a response to the foreseeable exhaustion of high-grade ore in the Mesabi Range. The 1950 distribution pattern was not much affected by the new situation with its heavier reliance on imported ores, but it is now apparent that the Atlantic Coast will regain some of its earlier position within the American steel industry. Several big plants have recently been opened[4] and others are under construction. No less than 40 per cent of the steel-making facilities under construction in 1952 were located in this area.[5]

The Atlantic Coast district is important also for refining of nonferrous metals. Baltimore is the largest copper-refining center in the world.[6] The copper rolling and drawing capacity is highly concentrated in the Connecticut valley which has about one-third of the national employment. The largest center in this area is Waterbury (B-type), known for its copper, brass and bronze products. Another copper town between this district and the Great Lakes region is Rome (A-type) in the Mohawk Valley.

The Alabama district is the largest outside of the Manufacturing Belt. The raw material situation for steel production is here almost perfect. Both a huge ore bed and vast deposits of excellent coking coal are available close to each other in the region. The low assembly costs for the raw materials are, however, counterbalanced by high transportation costs for the finished product when pig iron and steel are sold north of the Ohio River. As markets are thus restricted geographically, the expansion of this district is tied up with the general industrial development of the South

The western half of the United States, rich in nonferrous mineral production, is rather empty on the map of primary metal industries, at least in comparison with the northeastern part. Smelting is a major industry in several of the sparsely populated western states like Utah, Arizona, and Montana. It is of great significance for the economy of these states, even if it plays a rather subordinate role in the national pattern. Even in the West iron and steel plants employ more people than the nonferrous smelting and refining establishments. This area is well endowed with coal reserves. During World War II several of the many scattered iron ore bodies were investigated and it was found that there are considerable reserves of high-grade iron ore, particularly in southwestern Utah. Relatively low assembly costs for raw materials in Utah are offset, however, by high transportation costs for the finished steel.

Pueblo, Colorado, had for a long time the only big steel works in the West, specializing in such products as railroad equipment and barbed wire, which have a large market in this area. Pueblo gets coal from southern Colorado and iron ore from Utah. During World War II the tremendous shipbuilding program on the West Coast created a market for steel. Two integrated steel plants were built; one by the Federal Government at Geneva (near Provo) in Utah and the other by Kaiser at Fontana east of Los Angeles. The Geneva plant was sold to US Steel Corporation after the war. Both plants seem to have an expanding market in the rapidly growing Western states, especially in California.

Nonferrous smelting and refining has "created" only one city with more than ten thousand inhabitants in the West. Anaconda, Montana, grew up around the large copper smelter which was moved there from Butte, the mining town, because of the injurious fumes from its stacks. Great Falls, Montana, is an important copper

refining center. Salt Lake City and Tacoma have a large nonferrous smelting and refining production, chiefly of copper, but in these big cities the industry plays a rather subordinate role. Most copper smelters are located close to the mines but $2/3$ of the refining capacity is located on the East Coast, close to the market for refined copper.

The production of aluminum ingots in the Pacific Northwest, in Portland, Spokane and other cities, is the largest of any district in the United States,[1] but relatively few people are employed. The plants are located here because of cheap hydro-electric power. Bauxite is imported from South America and reduced to alumina in plants on the Gulf Coast. From there alumina is transported by railroad to the Pacific Northwest. The aluminum ingots are shipped from the Northwest, again by railroad, to rolling mills in the Manufacturing Belt and the Southeast.

It is a truism that distribution patterns are dynamic. A static picture, a map, is tied to a given date and might seem of limited value. On the other hand: changes of a magnitude to transform the general distribution of a mature industry like that of primary metal production (approximately = the iron and steel industry) in a mature industrial country like the United States do not take place overnight. A certain shift to the Atlantic Coast with its favorable location for the imports of Latin American, Canadian, African and Swedish ores has already been mentioned. The Western and the Southern districts have also expanded faster than the national production in the last two decades, but they still are rather unimportant. The area between the Mississippi and the Allegheny Mountains with its four-

1. After 1950 the center of the expanding aluminum industry has shifted to the Texas-Louisiana-Arkansas area. White and Foscue, *Regional Geography of Anglo-America*, New York 1954. Page 62.

# PRIMARY METALS

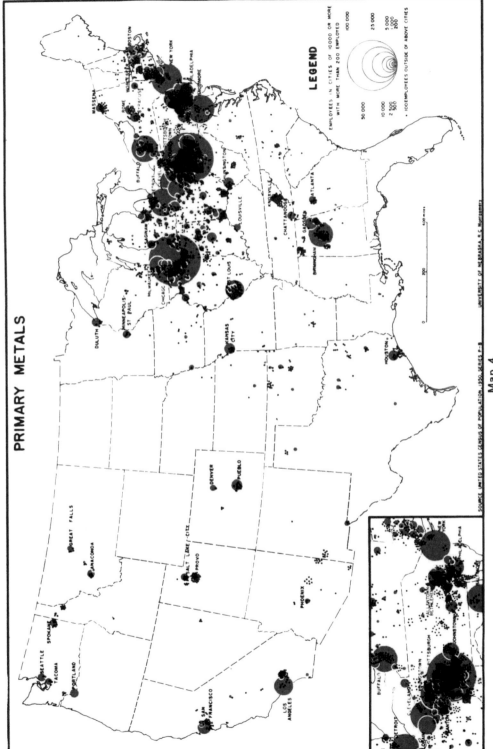

SOURCE: UNITED STATES CENSUS OF POPULATION, 1950, SERIES P-3    UNIVERSITY OF NEBRASKA, R.C. Montgomery

Map 4

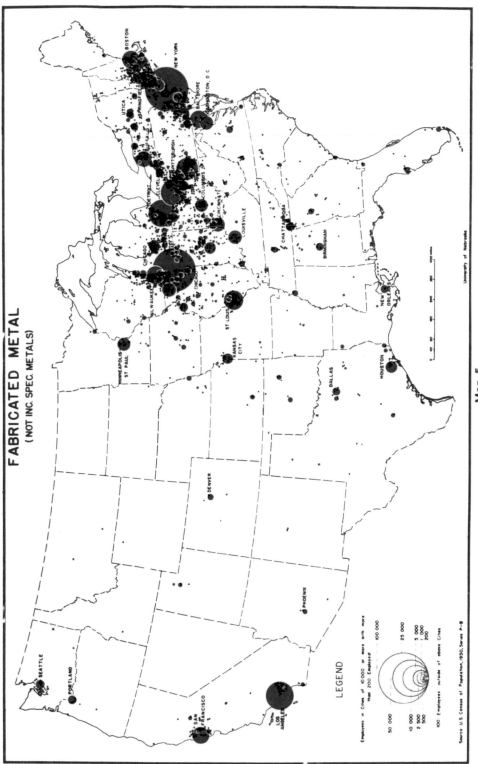

FABRICATED METAL
( NOT INC. SPEC. METALS)

LEGEND

Employees in Cities of 10 000 or more with more than 200 Employee

600 000

50 000

25 000

10 000

5 000

2 500

1 000

500

200

100 Employees outside of above Cities

Source: U.S. Census of Population, 1950, Series P-B

University of Nebraska

Map 5

fifths of the American smelting capacity has, however, four advantages:

1) The low-grade taconite ores, available in almost inexhaustible quantities in the Lake Superior area, can probably in the future be concentrated at costs which will make them competitive with high-grade ores from abroad. The big investments in beneficiation plants made by the ore companies point in this direction.

2) Construction of the St. Lawrence Seaway will allow big ore vessels to enter the Great Lakes and thus make imported ore available in the Lake ports at prices only slightly exceeding those paid on the Atlantic seaboard.

3) Canadian and American interests are developing high-grade ore deposits in two areas north of Lake Superior (Steep Rock, 140 miles west of Port Arthur, and Helen Mine, near Michipicoten).

4) Last but not least: as can be seen from Maps 5–9, the steel consuming industries are highly concentrated in the western and central parts of the Manufacturing Belt, where they exist in a symbiosis with the steel industry. A large scale migration of steel manufacturing can only be thought of in connection with a migration of the steel consuming industries. A geographic shift of steel manufacturing of a

1. *Tin cans*, 47 thousand employees (Ill. 12, Md. 5, Calif. 5); *Cutlery, handtools and hardware*, 154 thousand (New England 44, of which Conn. 27; Mich. 19, Ill. 18, N.Y. 17, Pa. 11); *Heating and plumbing equipment*, 151 thousand (Ohio 22, Ill. 16, Pa. 14, Wis. 13, Calif. 11, Mich. 11, Tenn. 6); *Structural metal products*, 212 thousand (Pa. 32, Ohio 23, N.Y. 20, Calif. 17, Ill. 13); *Metal stamping and coating*, 183 thousand (Ohio 27, Pa. 27, Mich. 26, Ill. 23, N.Y. 19, Wis. 9); *Lighting fixtures*, 48 thousand (Ill. 9, Ohio 8); *Fabricated wire products*, 61 thousand (Mich. 13, Ill. 9, Pa. 6); *Miscellaneous fabricated metal products*, 116 thousand (Ohio 25, Ill. 18, Pa. 13). Source: 1947 Census of Manufactures.

2. On manufacturing maps Baltimore usually is the southernmost big center in the Atlantic Coast region, but on Map 5 as well as on Map 15 of the printing industry the extreme service-city of Washington, D.C., stands out as an important center.

similar order as the one experienced in the American cotton industry therefore seems out of question even if, in the future, a substantial part of the ore supply should come from foreign mines.

## FABRICATED METAL PRODUCTS

The fabricated metal product industry manufactures the metals, primarily iron and steel, into a large variety of products which are neither machinery nor vehicles. The manufacture of machinery and vehicles is represented in the statistics as four separate groups of industries (Fig. 1). The present industry includes production of tin cans; of cutlery, hand tools and hardware; of heating and plumbing equipment; of structural metal products; of metal stamping and coating; of lighting fixtures; of fabricated wire products; and of other miscellaneous metal products.[1]

The distribution pattern of the fabricated metal industry is similar to that of primary metal production. Leading districts are the Atlantic Coast area from Boston to Washington, D.C.,[2] dominated by New York, the Pittsburgh-Detroit district and the Chicago region.

One-fourth of the employees in tin can manufacturing live in Illinois, which is in the center of an important canning region. Other leading canning states, such as California, Maryland, New Jersey, New York and Pennsylvania, are also important producers, each with about one-tenth of the national employment. Tin cans are made of steel with a very thin tin coating.

Production of cutlery, handtools and hardware, the type of goods which in Europe are associated with Sheffield, England, Solingen, Germany, and Eskilstuna, Sweden, is important in New England, especially in Connecticut, where the town of New Britain is a counterpart of these

Fabricated Metal Towns

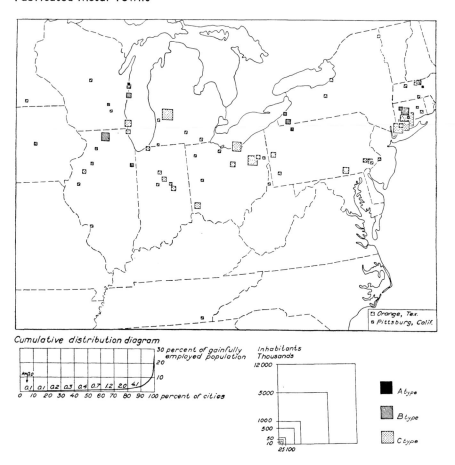

FIG. 5. Almost all of the fabricated metal towns are of B- and C-type.

European cities. Other leading states are Michigan, Illinois, New York and Pennsylvania. Rockford, Illinois, the largest town of B-type in this industry, has several hardware factories which have grown out from a lock factory, built in 1903 to supply the furniture industry, which was then the leading activity of Rockford, with locks and hinges, etc. Now the hardware industry is much more important than furniture manufacturing.[1]

The manufacture of heating and plumbing equipment is widely scattered over the continent, with Ohio, Illinois and Pennsylvania as the largest producers but with considerable production also in other states, among them California and Tennessee.

Structural metal products have a similarly wide manufacturing distribution, but the leading states are in the Manufacturing Belt, with the exception of California. Pennsylvania, Ohio and New York have the largest employment in this branch.

The metal stamping and coating industry is highly concentrated in the states of the Manufacturing Belt, where Ohio, Pennsylvania, Michigan, Illinois and New York all

1. John W. Alexander, "Rockford, Illinois: A medium-sized Manufacturing City," *Annals of the Association of American Geographers* (1953).

have more than ten per cent of the national employment. Production of lighting fixtures, fabricated wire products, and miscellaneous fabricated metal products are also largely confined to the same area with the same states leading in production.

MACHINERY

The mechanization of all industries and of housework has increased tremendously the output of each employee. The other side of mechanization is that a substantial and increasing sector of American manufacturing engages in the production of machinery. These firms are covered by two groups of statistics: electrical machinery industry and other machinery industry. They have a similar distribution pattern with a very marked concentration in the Manufacturing Belt.

On the whole, all branches of machine manufacturing[1] show a strong market orientation in their location. This is conspicuous for factories supplying specific, sporadic industries with machinery. Textile machines are manufactured in the old textile district of New England and the Middle Atlantic states; shoe machinery is made at Beverly on the outskirts of Boston and in the center of the most numerous localized group of shoe factories in the country,[2] oil-field machinery and tools are produced near the

oil fields in Texas, Oklahoma and California. Factories making machinery with a more ubiquitous demand, such as office machines and agricultural implements, generally have a favorable location with regard to the national market. The market orientation of the machine industry is of course not only a matter of minimizing transportation costs. Of great importance, but incalculable by any mathematical formula, is the personal contact between the men who produce the machines and those who use them in their factories. This contact gives impulses to inventions and improvements. It is facilitated if the two types of production are located close to each other.

Manufacturing of textile machinery did not move with the migrating cotton industry. The South is still dependent upon the North for most of its textile machinery.[3] A close parallel can be found in Northwest European countries, especially Great Britain. Machine manufacturing remained in the old textile centers when the cotton industry migrated to the former market countries—India and others.

Printing machinery is made in the two principal printing centers, New York and Chicago.[4] Oil field machinery and tools, as previously mentioned, are manufactured near the oil fields.[5] The machine tool industry used to be dominated by New England with its manufacturing tradition and skilled labor. The center of production has, however, moved to the leading metal manufacturing states of the Middle West. Ohio is the largest producer but also Michigan, Wisconsin and Illinois have a considerable output of these products, which are of strategic importance in all machine manufacturing.[6] Cincinnati is by far the largest single machine tool manufacturing center but also Cleveland, Detroit, Milwaukee and Chicago are large centers. Production of cutting tools, fixtures, etc. and of other metalwork-

1. Production of electrical machinery is dealt with in the next chapter.

2. E. M. Hoover, *Location Theory and the Shoe and Leather Industries* (Cambridge, 1937), p. 203.

3. Factories making *textile machinery* in 1947 employed 54 thousand people according to the Census of Manufactures (New England 33, of which Mass. 20, Me. 5, R.I. 4; Pa. 10; South 3).

4. *Printing machinery*, 25 thousand (N.Y. 8, N. J. 3, Ill. 5).

5. *Oil field machinery and tools*, 29 thousand (Tex. 13, Calif. 5, Okla. 4).

6. *Machine tools*, 71 thousand (New England 23, of which Conn. 8, R.I. 7, Mass. 4; Ohio 17, Mich. 8, Wis. 7, N.Y. 4, Ill. 4). The industry is of moderate size, although it has a key position in the manufacturing economy.

ing machinery, has a distribution pattern largely coinciding with that of the different metalworking industries, with a concentration in the Midwestern section of the Manufacturing Belt.[1]

The large-scale development of the farm machine industry is intimately connected with the settlement of the Middle West. This area, where land was plentiful but labor scarce, has in the last century witnessed tremendous advances in labor-saving devices. The American Middle West has in this respect played the same role in the economic history of the world as that played by the densely populated Northwest European countries in the development of land-saving devices, such as fertilizers, scientific breeding, weed control, etc. Several important farm implements were invented in the Middle West; the production of others was moved there from the East as it was early evident that the Middle West was going to play a leading role in farm mechanization. Cyrus McCormick constructed his reaper, which was to have a revolutionary effect on grain production, in the Shenandoah Valley of the Appalachian Mountains about 1831, but he built his first factory for large-scale production in Chicago (1847). John Deere designed a plow with a steel mold-board as a response to the need in the prairie region for a plow to break the prairie sod.[2] In 1855 he built a plant at his home town, Moline, Illinois, which is now, with contiguous Davenport and Rock Island, one of the largest centers for farm machinery in the world.

The competitive advantages of selling a full line of implements early favored the formation of big concerns selling in a nationwide market. The farm machine industry today is dominated by a few big companies which manufacture a large number of products, some of which are complicated and expensive (combines, cotton pickers, hay balers, etc.). A comparison of the actual distribution pattern of the American farm machine industry with Harris' maps of the market potential for farm implements[3] and of transport cost to the tractor market shows that the dominating Midwestern states, Illinois, Wisconsin, Iowa and Indiana,[4] also have the best market location in the nation.

As in the automotive industry, the big companies have assembly plants scattered over the continent to reduce distribution costs. Some types of farm machinery have a regionally concentrated market. Production of such machinery tends to be located with reference to the specific market area. A good example is International Harvester's plant at Memphis, Tennessee, which among other things produces cotton pickers. This plant was completed in 1948 but was conceived as early as in 1926,[5] which gives an idea of the careful planning that often precedes the establishment of new branch plants by big companies.[6]

The major farm implement centers in the United States are Chicago, Peoria, the Tri-Cities (Davenport, Rock Island, Moline) and Milwaukee. Peoria and the Tri-Cities are one-sided farm machine towns. All

1. *Cutting tools, fixtures etc.*, 89 thousand (Mich 24, Ohio 12, Ill. 11, Mass. 10, Conn. 6).
*Metalworking machinery, n.e.c.*, 55 thousand (Ohio 14, Ill. 8, Pa. 8, N.Y. 6).
2. "John Deere and Company," *Fortune*, May 1939.
3. Measured by the total number of tractors divided by transport cost. Chauncy D. Harris, "The Market as a Factor in the Localization of Industry in the United States." *Annals of the Association of American Geographers* (1954).
4. *Tractors*, 77 thousand (Ill. 39, Wis. 13, Iowa 10, Mich. 4).
*Other farm machinery*, 94 thousand (Ill. 29, Ind. 9, N.Y. 9, Iowa 7, Wis. 6).
5. C. E. Jarchow, Vice President and Comptroller, International Harvester Company, "Background of the New Harvester Plant at Memphis". Mimeograph, 1949.
6. The period between conception and completion was of course prolonged by the Depression and by Wold War II.

## Machine Manufacturing Towns

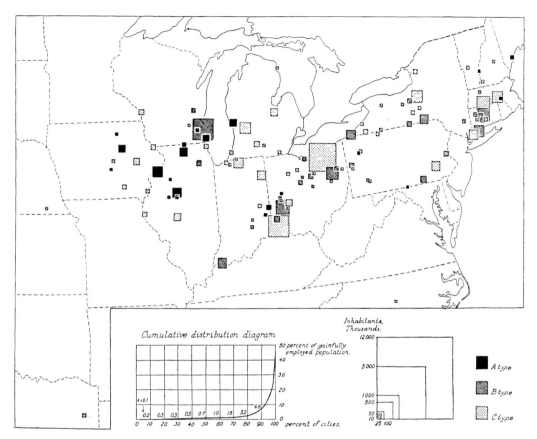

FIG. 6. The machine manufacturing towns are located in the Manufacturing Belt with a few (agricultural implement towns) in adjacent Iowa.

these, as well as several minor centers, are located in an area of the Middle West which, according to Harris, has the best market location. Within this area they lie in the eastern part, i.e., oriented towards the steel industry of Chicago or of cities farther east. The farm implement industry of the United States thus has become dominated by the area from which cheapest shipment to the total market can be made, but it has remained on the side of that area facing the source of materials and parts.[1]

The manufacturing of construction and mining machinery is dominated by the Manufacturing Belt.[2] Cities of the Lower Lake region, like Milwaukee, are favorably located for supplying heavy machinery for the iron ore mines in the Lake Superior region, cities in Ohio and Pennsylvania for making machines for the Appalachian coal mines. Construction machinery has a ubiquitous demand, but for firms with a national market the Manufacturing Belt will in most instances be the rational location, since it has the highest market potential. Chance factors will, however, operate to decide the specific location within this

1. Chauncy D. Harris, *op. cit.*, pp. 337ff.
2. *Construction and mining machinery*, 85 thousand (Ohio 17, Wis. 13, Ill. 10, Pa. 9).

general region—as in the case of elevator manufacturing at Bridgeport, Connecticut, or electric and steam shovel manufacturing at Marion, Indiana.[1]

Office and store machines[2] are made for the national market and the Manufacturing Belt should, according to Harris, be the logical location for their production. Even in this case there is a good correlation between actual distribution and optimum area delimited with Harris' method.[3] Dayton, Ohio, with the vast operations of National Cash Register, seems to be somewhat outside of the optimum area, especially since a large part of the output is exported.[4] A slight disadvantage in distribution costs can, however, be compensated by advantages of an early start, by favorable local combinations of a raw-material-producing steel industry, machine tool manufacturing, and a general industrial structure of the region favoring labor skill and efficiency. Among big office machine centers should also be mentioned Binghamton-Endicott, N.Y. (International Business Machines).

Service and household machine production is another industry located with reference to the total national market.[5] It is dominated by the Midwestern section of the Manufacturing Belt.

In conclusion: with proper allowance for the interdependence of locational factors and industrial distribution patterns, the market stands out as the single most important factor for the location of the machine manufacturing industry.

ELECTRICAL MACHINERY

The production of electrical machinery is more strongly concentrated in the two largest cities of the Manufacturing Belt, New York and Chicago, than is other machine manufacturing. Like several branches of the latter industry it is largely confined to the area best suited to serve the national market. As the two machine-manufacturing groups are thus very similar from a locational point of view, the electrical machine industry can be treated briefly here.

Almost from the beginning in the last decades of the nineteenth century the American electrical manufacturing industry has been dominated by two big companies, General Electric and Westinghouse.[6] The former company, the largest electrical manufacturing concern in the world, has its headquarters, research laboratories and largest manufacturing works in Schenectady, New York, which is an electrical machine town of A-type. The city grew very rapidly from 1886 with the establishment of the Edison Machine Works which became the largest General Electric plant when the latter was organized a few years later.

The main factories and headquarters of Westinghouse in Pittsburgh make electrical machine manufacturing the most important of the secondary metal products industries in the steel city. Both General Electric and Westinghouse have many plants scattered over the continent, chiefly in the Manufacturing Belt. Often the two big companies develop new products at their home plants but manufacture them elsewhere, frequently

1. Renner, Durand, White, and Gibson, *World Economic Geography* (New York, 1953).
2. *Computing and related machines*, 46 thousand (Ohio 14, N.Y. 13, Mich. 8).
*Typewriters*, 27 thousand (New England 13, N.Y. 13).
3. The market potential for points on the map is measured by the summation of retail sales in counties all over the country divided by transport cost from the points. Isopleths connect points with equal market potential. Harris, *op. cit.*, pp. 321 ff.
4. The export market was not considered in Harris' study, but it should further emphasize the great market potential of the New York region.
5. *Refrigeration machinery*, 129 thousand (Ohio 30, Ind. 21, Mich. 18, Pa. 18, N.Y. 6, Mass. 6).
*Domestic laundry equipment*, 28 thousand (Ill. 8, Iowa 5).
6. L. A. Osborne, "The Electrical Industry". *Representative Industries in the United States*, ed. by H. T. Warshow (New York, 1928).

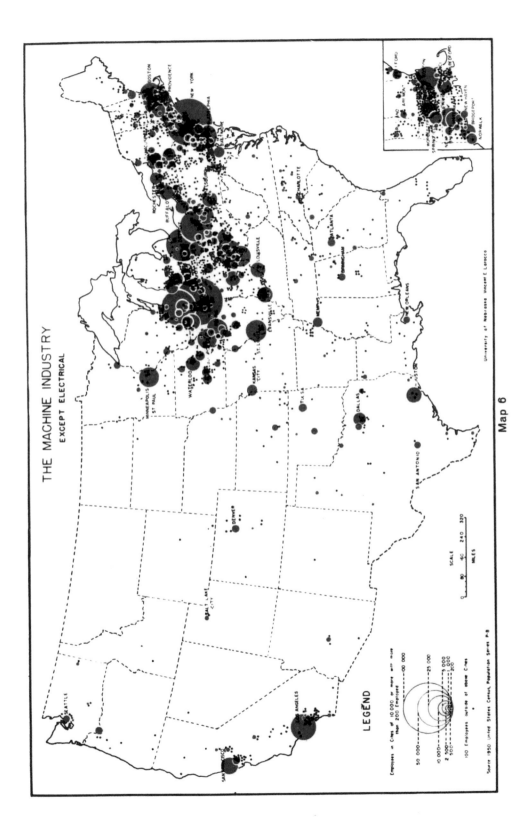

THE MACHINE INDUSTRY
EXCEPT ELECTRICAL

University of Nebraska   Vincent E. Lobosco

LEGEND

Employees in Cities of 10 000 or more with more than 200 Employed

100 000
50 000
25 000
10 000
5 000
2 500
1 000
500
200

100 Employees outside of above Cities

SCALE

0    80    160    240   320

MILES

Source: 1950 United States Census, Population Series P-B

Map 6

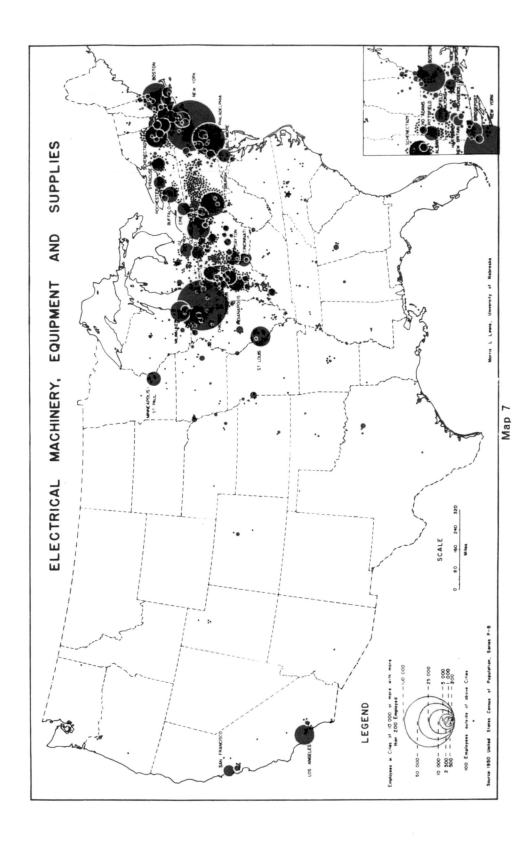

ELECTRICAL MACHINERY, EQUIPMENT AND SUPPLIES

LEGEND

Employees in Cities of 10,000 or more with more than 200 Employed

50,000
25,000
10,000
5,000
2,500
1,000
500
200

100 Employees outside of above Cities

Source 1950 United States Census of Population, Series P-8

SCALE

0    80    160    240    320

Miles

Morris L. Lewis, University of Nebraska

Map 7

Electrical Machinery Towns

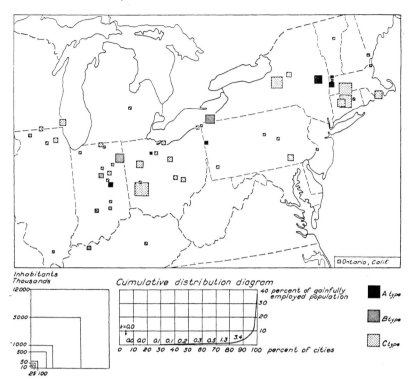

FIG. 7. The electrical machinery towns are highly concentrated in the Manufacturing Belt. The largest specialized city of this type is Schenectady, New York.

in smaller centers affording lower taxes, cheaper sites, lower labor costs, lower power costs, or a combination of these advantages. Thus, neither Schenectady nor Pittsburgh is an important center for the large production of electrical household appliances.[1]

MOTOR VEHICLES

Automotive manufacturing, in spite of its relative youth, has directly and, above all, indirectly made a greater impact on the American population geography than any other manufacturing industry. Mass produc-

tion of automobiles is only around 45 years old. Directly it has led to a tremendous urbanization in southern Michigan. Detroit is the most pronounced manufacturing city among the twelve American metropolises with more than a million inhabitants, and aside from Washington, D.C., it is the most one-sided of the big cities. Indirectly the automobile industry has had many effects. Among other things it has made possible a dispersion of urban settlement in all cities of the United States to an extent that is unknown in other parts of the world. About one-sixth of the national income is spent for automobiles and their operation. The automobile expenditures rank third in the family budget after food and shelter.[2]

Mass production of motor cars is inti-

1. G. E. McLaughlin, *Growth of American Manufacturing Areas* (Pittsburgh, 1938), p. 243.
2. E. L. Allen, *Economics of American Manufacturing* (New York, 1952), p. 287.

mately connected with the life of Henry Ford. There is some justification in fixing the birth of mass production techniques at 1913, when the crowning achievement, the continuously moving assembly line, was completed in Ford's new plant at Highland Park, Detroit. Before this date automobile manufacturing had gone through a rather long period of experimentation. The first motor cars were made in Germany (Daimler, Benz) about 1886, but the invention was first exploited commercially on a considerable scale in France, which remained the leading automobile country till 1906. In the meantime there had been several American inventors building motor vehicles since the 1890's, among them Henry Ford. During his early years as an automobile manufacturer Ford, like his competitors, made a variety of models at a variety of prices. Ford had designed eight models before his historic "Model T", which went into production 1908. He then concentrated on this model, which met with phenomenal success.[1] Production rose from 10,000 cars in 1909 to a peak of over two million in 1923. Ford's car, which remained substantially unchanged until 1927, when increasing competition forced him to design a new model, was intended as "a motor car for the great multitude," a *Volkswagen*, as the exponent of an equivalent German ambition thirty years later was appropriately called.[2]

The key to Ford's success was a progressive lowering of the price of his car, which meant a continuous increase in the number of potential customers. Between 1909 and 1922 the price was reduced from $950 to $295.[3] These price reductions were made possible by product standardization based upon the use of highly specialized machine tools, an intricate system of conveyor devices for transportation of parts within the plant, a minute subdivision of labor and an intensive supervision. Ford's production meth-

ods were as important as the car he built. The use of standardized, interchangeable parts and of the assembly line has become an indispensable constituent of modern mass production.

The American automobile industry is dominated by the "big three": General Motors, Ford and Chrysler. Together they produce 85 to 90 per cent of the total output.

General Motors was organized in 1908—the same year that Ford began production of his Model T—through a combination of several companies. It has never been a small firm, but it was surpassed by Ford within a few years. Not until 1927, when Ford closed down his plant and designed a new car, did General Motors again emerge as the leading automobile company with 40 per cent of the automobile business, a rate which it persistently holds.[4]

Chrysler is a relative newcomer in the field; the first Chrysler car was designed in 1924. From the beginning of the thirties Chrysler, after having bought other companies,[5] has held about the same share of the market as Ford.[6]

Automobile manufacturing is very sensitive to business fluctuations. The 1932 output was just one-fourth of the 1929 produc-

1. In 1912 Ford went one step further in his standardization and decreed that all cars would be ebonied with the same hue. "Any customer can have a car painted any color he wants so long as it is black." Allan Nevins, *Ford. The Times, the Man, the Company* (New York, 1954), p. 452.
2. Henry Ford, *My Life and Work* (London, 1923), p. 73.
3. E. B. Alderfer and H. E. Michl, *Economics of American Industry* (New York, 1950), p. 149.
4. General Motors Corporation produces the Chevrolet, Pontiac, Oldsmobile, Buick, and Cadillac passenger cars and the GMC truck and coach. General Motors is considerably more diversified than the other two companies. It also operates in other manufacturing industries, producing household appliances, locomotives, etc.
5. Chrysler produces the Plymouth, Dodge, De Soto, and Chrysler passenger cars and the Dodge truck.
6. Ford makes the Ford, Mercury and Lincoln passenger cars and the Ford truck.

tion. The first postwar depression in 1921 was also felt by the industry, as was the recession of 1938. During World War I the automobile plants became with part of their capacity subcontractors for shipbuilding and airplane manufacturing. This was repeated on a much larger scale during World War II. From 1942 to 1945 there were practically no automobiles made in the United States as plants were converted to the manufacture of munitions (airplanes and parts, tanks, bombs, machine guns, etc.).

The curve of passenger car registration (Fig. 8, B) shows the very rapid increase in the number of cars until 1929. The rate of increase levelled off somewhat during the twenties, but it was always considerably higher than the increase in population (curve A). In 1929 there was one car per 5.3 people. The upward trend was resumed after the minimum of 1933, and again the rate of increase was stronger than the population increase. In 1941 there was one car per 4.5 inhabitants. The war period resulted in a new decline, but after a low in 1944 and 1945 the curve pointed upward in a way that reminds one of the twenties. In 1951 there was one car per 3.6 persons.

"The family car" has been a reality in the United States for more than 25 years. The curve of car registration does not indicate any level of saturation, and logically there is none. Already social philosophers might speculate, however, whether the United States has a good balance between the proportion of the national income spent for a high transportation standard and the proportions spent in other ways, e.g. for housing. Seen from the point of view of the individual, the automobile is a necessity in the United States. Cities are planned and built for car owners: distances in the urban agglomerations are long and public transportation is poor.

The American automotive industry is very

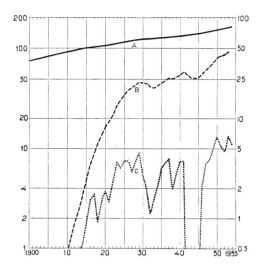

FIG. 8. Population development (A), car registration (B) and production of cars (C) in the United States. The scale to the left refers to population, the one to the right to cars.

strongly concentrated geographically. Detroit dominates the industry in a way that has an equivalent only in New York's dominance of apparel manufacturing. Mass produced motor vehicles are just the type of products for which a marked geographic concentration can *a priori* be expected. In this industry very big production units are more efficient than smaller ones. If one point on the map has a slight advantage as a location for one of these large plants, it will be favorable also for the others. There are no serious drawbacks in being located close to the competitors, which all have the nation as their market, but rather undeniable advantages. Such a location makes available skilled labor and technicians which are to be found in any specialized city. The interest of the region is more or less geared to the industry (education, banking, etc.), and there develops a net of subcontractors and firms producing services especially for this industry.

The disadvantages of such a concentration of production, which usually leads to a one-sided economy in the city and region, are of social and military rather than economic nature. In a general economic depression or in the case of stagnation in the industry, the city will be seriously hit. The military strategists want to have manufacturing dispersed instead of concentrated, since such key industries as automotive manufacturing are bomb targets of high priority.

The increases in distribution costs of the bulky product, which result from a geographic concentration of production, are not large enough to offset the savings in production costs, especially as they can be reduced by a system of well located assembly plants to which parts can be shipped from the manufacturing plants at a fraction of the cost for transportation of the assembled product. Even before World War I, Ford had assemblies in big cities all over the United States and this system has later been followed by other companies.

A geographic concentration of the automotive industry would probably occur even in a hypothetic case with all points on the map equally located with regard to raw materials and markets. In reality raw materials are very unevenly distributed and the same is true about population (the market). The location of steel manufacturing in the United States has been dealt with in an earlier chapter. For a heavy steel consuming industry like automotive manufacturing a central location with regard to the steel industry is essential.[1] As the center of gravity of the American population falls within the Lower Lake region, this area is a strategic location also with respect to markets. The situation can be summed up in this way: given the distribution of coal and iron ore in the United States and the steel industry built on these resources, and given also the population distribution, all points outside of the Lower Lake region will be at a disadvantage as centers for auto-

motive production compared with points within this region. The Lower Lake area is the logical location for the automobile industry.

But why is Detroit the world's leading car city and not Cleveland or Chicago or any other city on the lakes with a general location as favorable for automobile production as that of Detroit? The answer is one which is often found to similar questions: Detroit had the advantage of an early start. The cities of southern Michigan, originally founded as trading posts, usually had a period during the nineteenth century when lumbering was their dominant industry, and this in turn was followed by manufacturing based on wood and later also on iron and steel. Detroit, Flint, Lansing and Pontiac were known for their wagons, carts, buggies and carriages. Grand Rapids became the American furniture city. Around the turn of the century Detroit was the biggest city in Michigan, with over 200,000 inhabitants. It was a leading manufacturing center for gas engines, carriages, and wagons and it was well equipped with machine shops. In Europe and in the United States a large number of people were tinkering with self-propelled vehicles as early as the 1890's. It was hardly an accident in the statistical sense that several of these men were living or had got their apprenticeship in Detroit, with its vehicle and motor milieu. Detroit was well suited to become the automobile city. The leading car factories in Detroit during the first years of the century—Cadillac, Packard, Wayne, Oldsmobile and Ford—early acquired national fame. The birth rate as well as the death rate was high for automobile factories during the pioneer years. There was also a certain migration of firms, mainly to

1. An automobile requires around 1 $^3/_4$ tons of steel (*Steel Facts*, October 1953). The automotive manufacturing is the largest steel consuming industry in the United States with an estimated 16.8 per cent of the total production (1948).

## Automotive Manufacturing Towns

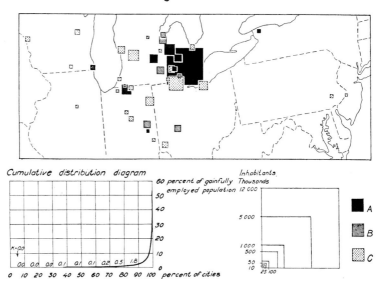

FIG. 9. Detroit is the largest automotive manufacturing town in the world. Flint, Lansing, and Pontiac, Michigan, and South Bend, Indiana, also rank among the world's leading automotive manufacturing centers.

Detroit, from even rather distant places in the Manufacturing Belt, but also in a few cases from the city itself to nearby towns. Detroit had established its position as the leading American motor car center several years before Ford with his Model T emerged as the giant of the field.[1]

The automobile factories in the extensive urbanized area of Detroit, with its 2.6 million inhabitants, are widely scattered. Ford's River Rouge Plant at Dearborn, 10 miles from downtown Detroit and near the farm on which Henry Ford was born, is a highly integrated unit, with steel plants, a glass plant, etc., in addition to the car manufacturing facilities. It employs about 90,000 people. Usually the automobile plants are of a more moderate size. General Motors has several units in Detroit ranging from 5,000 to 10,000 each. Chrysler has the big Dodge Plant with 25,000 employees, as well as smaller units. In addition to the automobile factories proper there are a large number of integrated or independent plants producing parts.[2]

Besides Detroit southern Michigan has four automobile cities of A-type, Flint, Lansing, Pontiac, Ypsilanti (including Willow Run), and several towns of B- and C-type. Southern Michigan would dominate the world's automobile manufacturing even without Detroit! Flint is the most specialized of all American cities dominated by the automotive industry; 53 per cent of its gainfully employed population are engaged in automobile manufacturing. It is second only to Detroit among the automobile centers of the world. Flint is known as the

1. In 1900 about 4,000 American cars were manufactured, of which steam and electric models comprised more than three-fourths. About 50 per cent of the cars were made in New England. However, as less than one-fifth of the gasoline cars originated here and the steam and electric models were soon to disappear from the field, the early prominence of New England was of little relevance for the final location of the industry. Detroit was the leading gasoline-car center from the beginning.

2. Clyde F. Kohn, "Detroit", *Industrial Cities Excursion, Guidebook.* XVIIth International Geographical Congress (Washington, 1952).

home town of W. C. Durant, who after having made himself a fortune as a carriage manufacturer made a success of the Buick company, which became a cornerstone in the General Motors Corporation when it was formed in 1908 under Durant's leadership. Lansing was the home town of one of the American automobile pioneers, R. E. Olds, who was a mechanic in his father's machine shop. Finding financial backing in Detroit, Olds started the production of his Oldsmobile here in 1901 but when a new big plant was built in 1904 it was located at Lansing.

As can be seen from Map 8, many employees of the automobile industry live outside of the cities. Many are commuters, but a considerable number are employed in small, decentralized plants producing automobile parts. Ford has, for instance, fifteen plants ranging in size from 25 employees to 400 within a radius of fifty miles from the River Rouge plant.[1]

The automobile towns outside of Michigan are well concentrated in adjacent states (Fig. 9). South Bend (Studebaker) and New Castle, Indiana, Kenosha (Nash), Wisconsin, and Lockport, New York, are automobile towns of A-type. There are also a number of B- and C-towns. The large cities Cleveland, Toledo, Buffalo, Chicago and Milwaukee are all important automotive manufacturing centers. Part of the capacity in these cities and practically all in more remote big cities is regional assembly plants. From these the assembled cars are transported on specially constructed trucks, a conspicuous vehicle seen on all main roads in the United States, to the local dealers.

TRANSPORTATION EQUIPMENT OTHER THAN MOTOR VEHICLES

The transportation equipment industry differs considerably from the other branches of

metal manufacturing in its distribution pattern. The Manufacturing Belt does not dominate on Map 9 even if the Atlantic Seaboard is very important. The Pacific Coast has four centers of heavy concentration, and such inland states as Texas and Kansas rank high. The industry is made up of three main branches with different distributions: shipbuilding, airplane manufacturing and production or railroad rolling stock.

*Shipbuilding*

American shipbuilding in peacetime is of minor significance. During the first decades of the last century, or up to the Civil War, the United States was, however, the leading shipbuilding nation, surpassing even England. Yankee clipper ships, built in New England, represented the apex in wooden ship design. New England's important whaling industry required a large number of vessels. The American shipyards could draw on an ample supply of suitable timber, whereas the domestic lumber basis for the English industry was limited indeed. After the Civil War the United States rapidly declined to a position of minor importance among the shipbuilding nations. Of several concurring reasons behind this change the following are probably the most important. England was several decades ahead of the United States in the development of its iron and steel industry. When steam ships displaced wooden sailing vessels England rapidly took the lead in the new shipbuilding technology. In time these changes coincided with the American boom in railway construction, with the opening up of vast areas in the interior for agriculture and with the tremendous industrialization in the Manufacturing Belt. The available capital, of which much was borrowed from Europe, could be engaged more profitably in other

1. Alfred J. Wright, *United States and Canada* (New York, 1948), pp. 292 ff.

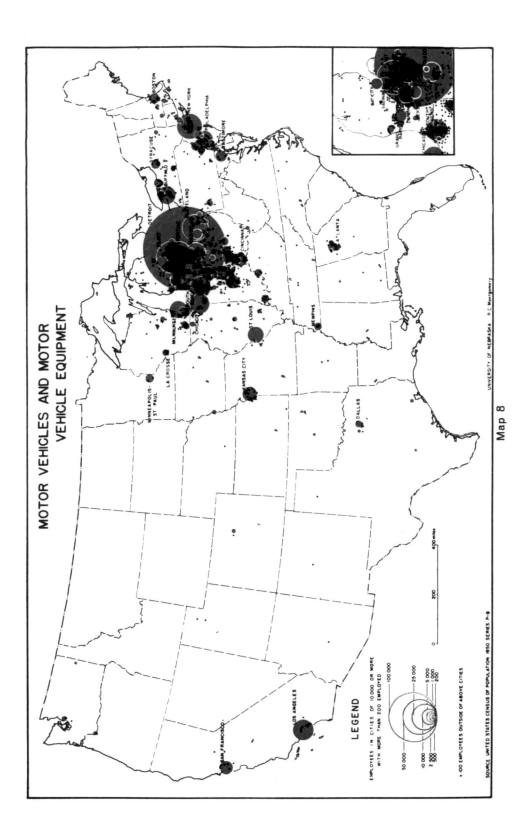

# MOTOR VEHICLES AND MOTOR VEHICLE EQUIPMENT

## LEGEND

EMPLOYEES IN CITIES OF 10 000 OR MORE
WITH MORE THAN 200 EMPLOYED

100 000

50 000

25 000

10 000

5 000

2 000
1 000
200

• 100 EMPLOYEES OUTSIDE OF ABOVE CITIES

0          200          400 miles

SOURCE: UNITED STATES CENSUS OF POPULATION 1950 SERIES P-B          UNIVERSITY OF NEBRASKA   R C Montgomery

Map 8

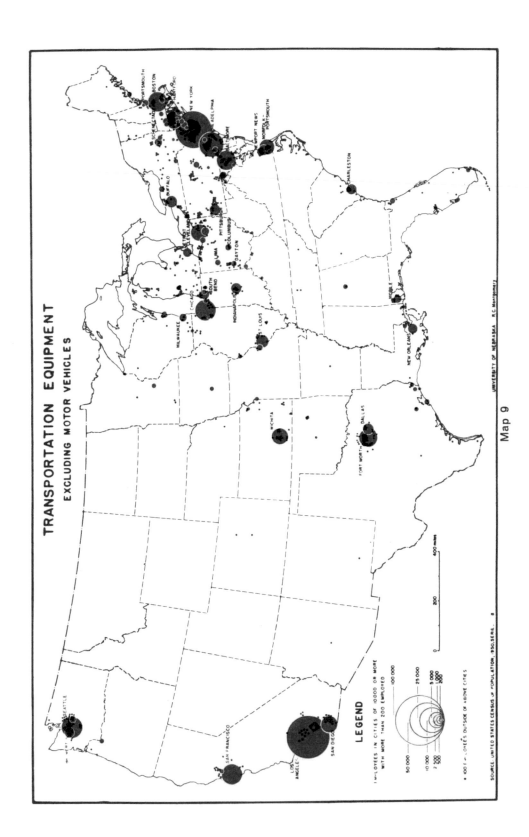

TRANSPORTATION EQUIPMENT

EXCLUDING MOTOR VEHICLES

LEGEND

EMPLOYEES IN CITIES OF 10,000 OR MORE
WITH MORE THAN 200 EMPLOYED

100,000

50,000

25,000

10,000

5,000

2,000
1,000

• 100 EMPLOYEES OUTSIDE OF ABOVE CITIES

SOURCE: UNITED STATES CENSUS OF POPULATION, 1950, SER. H.   B

UNIVERSITY OF NEBRASKA   R. C. Montgomery

Map 9

activities than shipbuilding, in which the British had many advantages.[1]

At the beginning of the First World War about 90 per cent of American foreign trade was carried on foreign ships, and there were only a few shipyards in operation in the United States. During the war the technique of mass production was applied to shipbuilding, and the American production was stepped up to a peak output of five million dead-weight tons, which should be compared with an annual production of 0.1 to 0.2 million tons before and after the war. During the Second World War the same story was repeated on a much larger scale: the year 1943 saw an American production of 18.5 million tons. Shipbuilding became a leading American industry. The shipyards were turned into assembly plants and standardized parts were shipped in from steel mills and metal manufacturing plants, often located far inland. There was no question of custom-building the ships as is normal in peacetime, when quantity was the most important matter. The U.S. Government adopted the new welding technique for their standard types of ships, thus permitting a more rapid deployment of the country's shipbuilding resources and much greater production than would have been possible with riveted construction.[2]

Several coastal regions took part in this inflated war production of ships. On the Gulf Coast shipyards were in operation at Tampa, Mobile, Beaumont, New Orleans, Houston and other places. Nearness to steel mills at Birmingham and Houston made this coast a rather favorable location. It was less exposed to attack than the other coasts. Pascagoula, Mississippi, is a shipbuilding town of B-type.

On the Pacific Coast most of the shipbuilding activity was concentrated in the San Francisco Bay region and the Puget Sound area, both of which have some shipbuilding activity even in peacetime. The United States Navy has shipyards both in the San Francisco region and at Bremerton, west of Seattle. In Bremerton, a major naval repair base, the population trebled during the war. Seattle has shipyards, but airplane manufacturing accounts for most of the city's prominence within the transportation equipment industry. Napa, near San Francisco, is a shipbuilding town of B-type.

A large share of war production took place on the Atlantic Coast, normally the most important shipbuilding region in the United States. The Delaware River, with yards at Philadelphia and Wilmington, is sometimes referred to as "the American Clyde." Shipyards are also located in the Baltimore and New York regions. On the Cheasapeake Bay Newport News, with one of the largest shipbuilding yards in the world, and Norfolk - Portsmouth, headquarters of the Atlantic Fleet and site of the Norfolk Navy Yard, are important shipbuilding centers even in peacetime. Further south on the coast Charleston stands out as a shipbuilding city of B-type. Of the once-flourishing shipbuilding activity in New England only traces can now be found. The small city of Bath, Maine, has been a shipbuilding center since the eighteenth century. Bath has boomed during the wars and declined in peace time. It is one of the two one-sided shipbuilding cities in the United States. Another city with a long shipbuilding tradition is Portsmouth, New Hampshire. The shipyards in Boston are rather important but play a subordinate role in the economy of the city. Aircraft manufacturing is also represented in Boston.

The shipbuilding activity on the Great

1. E. B. Alderfer and H. E. Michl, *Economics of American Industry* (New York, 1950), pp. 138 ff.
2. Lloyd's Register of Shipping. Annual Report 1952.

Lakes is mainly concentrated in the big cities. There is only one town in which the yards are of great relative importance: Manitowoc, Wisconsin.

## Aircraft

Aircraft manufacturing[1] is essentially an assembly industry. Airframes, engines and propellers are seldom produced at the same plant. The final assembly plants obtain from other sources completely assembled engines, propellers and instruments. The situation is similar in the engine and propeller branches of the industry. Even if thousands of plants contribute parts to the airplanes through subcontracting, the aircraft industry proper is confined to the relatively few factories making the final assembly of airframes, engines and propellers.

The first boom in aircraft manufacturing came with the First World War. Within twelve months world output increased from ten planes a week to nearly 500. At the time of the Armistice production in the United States had reached a rate of 20,000 planes a year and 200,000 people were employed. The total American war production was 16,000 planes. The industry was concentrated geographically. The most important centers were Dayton, home town of the Wright brothers, Buffalo, Detroit and New York. Engine production was still more concentrated. Over one-half of the total value came from Detroit, one-fourth from New York.

At the termination of the war airplane production collapsed. A second, more moderate boom came in the late 1920's with the great stimulus provided by Charles Lindbergh's spectacular flight in 1927 between New York and Paris, which more than any other single event seems to have made the public air-conscious. The depression in the early thirties hit the aircraft industry harder than it did manufacturing in

general. But worldwide re-armament from about 1935 led to a rapid increase in employment. Besides the growing military orders—military planes make up the bulk of business in the industry with normally 75 to 90 per cent of the total value—there was an increasing market for passenger planes. Commercial craft were sold to Europe, where military deliveries had priority in domestic aircraft production. Simultaneously passenger travel in the United States increased with improved comfort, safety and regularity of air travel and with reduction of fares.

During the twenties and thirties there was a continued westward movement in the industry and a tendency towards concentration. Particularly noteworthy was the heavy concentration in Los Angeles and San Diego, but also in Seattle, with the largest company on the West Coast (Boeing)[2] and Wichita, Kansas, with a number of companies manufacturing small planes. Practically all of the larger craft, both transports and heavy bombers, were produced by Pacific Coast plants even before World War II.

Most of the shift in the aircraft industry has been in the airframe branch or the final assembly of planes. The engine and propeller plants remained at or near their original sites. New York (Paterson), Hartford and Indianapolis supplied all the engines, and Hartford 90 per cent of all propellers. The engine and propeller branches of the industry were well entrenched in the metal manufacturing Northeast with

1. This chapter draws heavily on W. E. Cunningham, *The Aircraft Industry, A Study in Industrial Location.* (Los Angeles, 1951).
2. William E. Boeing, connected with timber and mining operations in the Pacific Northwest, was interested in flying as a hobby. During the early period of World War I he had an accident with his plane, which he repaired himself. He became so interested that he decided to build a new plane. From this accident arose one of the world's leading aircraft companies, making mainly heavy bombers and large commercial transports.

its vast store of skilled labor and its multitude of subcontractors.[1] The producers of
airframes, on the other hand, were induced
by the mild climate and favorable flying
weather of the Pacific Southwest.

In spite of the rapid increase, aircraft
manufacturing in 1939 still was a relatively
insignificant industry. In terms of employment it was about the size of the men's and
boys' shirt industry. The motor vehicle
industry turned out twenty-two times as
much in value.

The Second World War had the same
effect upon airplane manufacturing as it had
upon shipbuilding. In a short time the aircraft industry grew from a minor position
in American industry to become the largest
in terms of employment and value of output.
At the peak period, November 1943, 2.1
million people were employed, and the
twelve largest companies each engaged more
wage earners than the whole industry in
1939. When President Roosevelt in a mes-
sage to Congress in the spring of 1940
expressed the need for 50,000 planes a year,
the industry was geared to an annual production of about 5,000 planes. The volume
mentioned by the President was reached as
early as 1942, and the peak year 1944 saw a
production of 96,000 planes. The huge
expansion was attained by methods similar
to those applied in shipbuilding during the
same period. Existing plants were enlarged,
new plants were built, a large portion of the
automobile industry was set aside for production of aircraft parts, work was divided

into simple tasks so that new workers could
be trained in short time, and much work
was done by subcontractors.

The ephemeral distribution pattern of the
war period will not be dealt with in detail
here.[2] After the war, came some lean years
for the aircraft industry, and in 1947 the
companies were struggling for their existence. Large military orders in 1948 and
later have led to a revival. The location of
the industry in 1950 can be described as essentially a return to the prewar pattern.

The only one-sided aircraft manufacturing city in the United States, Renton, southeast of Seattle, is dominated by a huge
Boeing plant.

Of the B-type cities, Wichita, Kansas, has
already been mentioned. Wichita is a center
for airplane manufacturing of long standing,
and it is known for its production of small
planes, which have found a market especially
in the Midwest and the West. The prairie
region offers natural advantages for test
flights, but the growth of Wichita as an airplane center is mainly the result of initiative
by local men, backed by capital accumulated
in the oil business. After the war Wichita's
position as an aircraft manufacturing center
was strengthened by the strategically motivated transfer of part of Boeing's production
of bombers from Seattle.

Fort Worth, Texas, is another aircraft
manufacturing city of B-type in the prairie
region. The largest factory in Fort Worth
makes heavy bombers. World War II saw
a rapid growth of airplane manufacturing in
Fort Worth and Dallas, with their good
flying weather and their proximity to large
military airbases. After the war production
declined, but the Dallas-Forth Worth
region still is important in the final assembly
of airplanes.

Hagerstown, Maryland, is the only city of
B-type in the eastern United States. It is
known for final assembly of aircraft. Nearby

1. Subcontracting to the extent of 10 per cent
was typical of the industry in 1940. In December
1944 airframe manufacturers subcontracted 38 per
cent of their production and engine manufacturers
28 per cent.

2. The military strategists advocated a location of
new aircraft producing facilities between the Appalachians and the Rocky Mountains, out of reach for
carrier borne enemy planes. As a result, the relative
importance of the Pacific Coast was considerably
smaller during the war years than before and after
the war.

Baltimore also has a large airplane production.

Of the two outstanding production centers in southern California, San Diego is an aircraft manufacturing town of C-type. In Los Angeles aircraft production is the leading manufacturing industry, engaging more than four per cent of the gainfully employed urban population, just below the limit, which would have made Los Angeles a C-type city. As has been mentioned, Los Angeles and San Diego specialize in the final assembly of planes, in which branch they dominate the American production. The mild climate of southern California appears to have been the main factor attracting the aircraft industry to this region. Weather permits test flights all the year round; much work on the big planes with their huge wing span can be carried on out-of-doors and parts and equipment can be stored in the same way; the mild winters reduce heating needs and thereby construction costs, which is an important consideration where hangars with millions of square feet of floor space are involved.

Hartford, Connecticut, especially known as a world center for propeller production, is an aircraft city of C-type. Engines and propellers are still made almost exclusively in the Manufacturing Belt with its diversified metal manufacturing structure.

## Railway Rolling Stock

Railway rolling stock production is located in the Manufacturing Belt. Proximity to the steel industry and availability of labor trained in metal manufacturing make this general area the logical location for such heavy metal production as that of railway cars and locomotives. The industry became firmly rooted in the Manufacturing Belt during its expansion period, the later half of the last century, but especially after the Civil War, when production of railway

equipment was of greater relative importance in the American economy than it is at present. Light metals have recently made inroads on steel in this industry, but there has been no tendency to relocation as the established plants accepted the new materials.

Chicago is the most important manufacturing center for railway cars. This city is both the greatest hub in the American

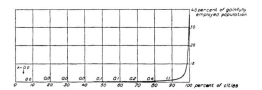

FIG. 10. Cumulative distribution diagram for transportation equipment other than motor vehicles.

railroad net, with a traffic exceeding that of St. Louis and New York combined, and the largest and most diversified metal manufacturing center with a production of basic steel as large as that of Pittsburgh. The former town of Pullman, now a suburb of Chicago, was built as a "model town" in 1881 by George M. Pullman, originator of the railway sleeping car, for employees of his new plant, the largest passenger car factory in the world. Neighboring Michigan City, Indiana, is a C-type center making Pullman cars.

The other large Midwestern railroad junction, St. Louis, has a considerable production of railway equipment. As can be seen from Map 4, it has a steel industry of its own. Nearby St. Charles, Missouri, is a railroad car city of B-type.

Berwick, Pennsylvania, ranks high in the manufacture of freight cars. It is the only one-sided town dominated by the railway rolling stock industry.

Philadelphia has for decades been the world's leading producer of locomotives.

The advantages of Philadelphia as a center for the manufacture of locomotives and other heavy steel equipment have been strengthened recently with the completion of huge steelworks on the Delaware River, based on imported ores. Other important centers of locomotive production are Chicago, Pittsburgh and Schenectady, New York. During the last decade there has been a rapid change to Diesel-electric locomotives, but steam engines still dominate. Only a small fraction of the American railroad net is electrified.

OTHER DURABLE GOODS

The first of the two groups into which American statistics collect manufacturing branches which do not fit into any of the specified groups embraces such incommensurable industries as cement manufacturing and the manufacturing of photographic equipment. The branches included, however, are sporadic in their occurrence, and it will be possible, except for the biggest cities where several or all branches may be represented, to determine the city-forming branch of any specified city with the help of *The Columbia Lippincott Gazetteer of the World, Webster's Geographical Dictionary* or any of the common encyclopedias.

*Cement*

The cement industry[1] in the United States for all practical purposes is identical with the manufacturing of Portland cement. The older types, natural cement and pozzuolana, constitute less than 1 per cent of the total output. Although Portland cement was first made in 1824 by Joseph Aspdin, a bricklayer of Leeds, England, the first American mill

was not built until 1872, about 25 years after the development of commercial production in Europe. Not until the 1890's did domestic production of Portland cement exceed imports. A very fast increase in production started about 1900 with the introduction of the rotary kiln.[2] This and other inventions and improvements made cement manufacturing one of the most highly mechanized of American industries. The standardized product could be sold at decreasing prices in an expanding market. The greater heights of buildings required stronger and more durable materials and the automobiles created a demand for more and better roads. The extensive mechanization of cement manufacturing makes it a rather small industry from the point of view of employment.

During its early years the Portland cement industry was heavily concentrated in the Lehigh Valley of eastern Pennsylvania. Deposits of exellent limestone, access to anthracite coal and nearness to a big market made this the dominating cement producing area with about 70 per cent of the 1900 production. The Lehigh Valley still is the leading cement-producing region of the United States, but now with less than 20 per cent of the national output in spite of a considerable increase in actual tonnage. Allentown, Easton and Phillipsburgh (N.J.) are the leading centers of cement production in this region.

American cement plants now are scattered in rough proportion to the population distribution. Cement has a low value per unit of weight, which means that distribution costs are high. Since about two tons of raw materials (predominantly limestone) and fuel are required to produce one ton of cement, the mill must be located at the source of limestone or in a port. The widespread distribution of limestone and of cheap fuels makes it possible to locate the

1. *Hydraulic cement manufacturing* in 1947 employed 36 thousand people according to the Census of Manufactures (Pa. 7, Calif. 3).
2. Alderfer and Michl, *Economics of American Industry* (New York, 1950).

cement mills with primary consideration given to the market.

## Clay-working Industries

The clay-working industries have the raw material in common but their products are highly diversified and their distribution patterns rather different.[1]

Brick and tile manufacturing is scattered in a rough proportion to the population. The products are bulky and cheap relative to weight; the raw material (clay) and the fuel are more or less ubiquitous. The brick plants thus will be located at the source of raw material but close to the market. It is estimated that 90 per cent of the bricks are sold within a radius of 67 miles from the plant.[2]

The clay refractories or the heat-resisting products industry is located principally in the Allegheny Plateau of Pennsylvania and Ohio, where the fire clay is interlayered with bituminous coal. The steel industry is the chief market. Another concentration of this branch is in Missouri.

The pottery industry produces items with a rather high unit value. They can be marketed over a wide area. Pottery products even enter international trade. It is one of the industries in which foreign competition (United Kingdom, Japan) has been felt most strongly. Like other clay-working industries pottery fabrication is a high labor cost industry: it has one of the highest ratios of wages to value of product found in any industry.[3] American pottery manufacturing has been able to expand chiefly because of a high tariff wall.

In its early phase the industry was heavily concentrated in Trenton, New Jersey—still important, especially for sanitary ware—but from the end of last century East Liverpool, Ohio, grew rapidly and for several decades has been the American pottery center. The Upper Ohio Valley district

of Ohio, West Virginia, and Pennsylvania—centered on East Liverpool and embracing a large portion of the American pottery industry—is very well located with respect to fuel, both coal and natural gas, and it also has a good market location for products sold all over the continent. Clay is procured from several different states and even from abroad (Cornwall, England), indicating a relatively unimportant role for the raw material as a locational factor in pottery manufacturing.

## Glass

The glass industry[4] comprises three distinct branches with the same fundamental technology but with different products and markets. They make pressed and blown glassware, glass containers and flat glass.

The most important of the raw materials is sand. The industry can easily be supplied with its requirements of suitable sand from deposits in widely scattered states. When 60 to 70 per cent of the domestic supply comes from Illinois, Pennsylvania and West Virginia, it is chiefly because the deposits of these states are relatively close to the glass plants which have been located with primary

1. A few of the biggest branches according to the 1947 Census of Manufactures:
   *Brick and hollow tile*, 30 thousand employees (Pa. 3, Ohio 3).
   *Clay refractories*, 18 thousand (Pa. 5, Mo. 5, Ohio 3).
   *Vitreous-china food utensils*, 11 thousand (Pa. 4, Ohio 2, W. Va. 1).
   *Earthenware food utensils*, 17 thousand (W.Va. 6, Ohio 5, Pa. 1).
   *Porcelain electrical supplies*, 12 thousand (N.Y. 2, N.J. 2, Ohio 2).
2. Alderfer and Michl, *Economics of American Industry* (New York, 1950).
3. *Ibid.*, table 4. Based upon the 1947 Census of Manufactures.
4. *Flat glass manufacturing* in 1947 employed 27 thousand people according to the Census of Manufactures (Pa. 10, Ohio 5, W.Va. 4).
   *Glass containers*, 47 thousand (Ohio 10, Pa. 8, W.Va. 8, N.J. 6).
   *Pressed and blown glassware, n.e.c.*, 42 thousand (Pa. 10, Ohio 10, W.Va. 8, N.Y. 7).
   *Products of purchased glass*, 22 thousand (N.Y. 5).

## Other Durable Goods Manufacturing Towns

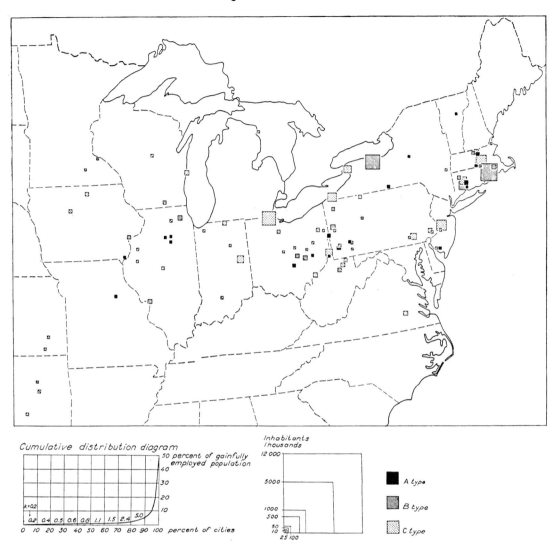

FIG. 11. Rochester, New York, is the American photographic equipment town, Providence, Rhode Island, is known for its silverware and jewelry, and the many small cities in western Pennsylvania, West Virginia, and Ohio are primarily glass and pottery towns. Toledo, Ohio, is also a glass town.

consideration being given to fuels and markets.[1] Within the wide area having a favorable market location, the industry seems always to have chosen sites with access to cheap fuel: first charcoal, later coal and

1. E. B. Alderfer & H. E. Michl, *Economics of American Industry* (New York, 1950).
2. E. M. Hoover, *The Location of Economic Activity* (New York, 1948).

finally natural gas. Improvements in the utilization and transport of fuel have reduced the importance of fuel as a locational factor in general, but glass manufacturing, requiring immense amounts of heat to melt the raw materials, is one of the few industries still significantly influenced by geographic differences in fuel costs.[2] All three branches of the industry are highly concentrated in Pennsylvania, Ohio and West Virginia, especially in the Appalachian Plateau section of these states. The Appalachian glass region also stretches into western New York.

The flat glass branch is characterized by very large plants, among the largest of any manufacturing industry. The other two branches also have plants of considerably above-average size. The economy of large-scale production partly explains why the glass industry is geographically concentrated in spite of its bulky and relatively cheap products made from raw materials and fuels, available in scattered localities all over the continent.

In the charcoal era Philadelphia and several small towns in southern New Jersey were among the leading American glass centers. The forests on the sandy soils of south central New Jersey seem to have been the chief attraction. In this area Millville and Bridgeton still are glass cities. The industry has survived here partly because of a favorable market location, partly because of specialization in laboratory glass and other quality products.[1]

Soon after the Civil War, with the adoption of coal as a fuel, the center of glass making moved from New England and the Eastern states to the Upper Ohio Valley. Pittsburgh became the leading glass town. When the industry later turned to natural gas, an ideal fuel in glass making, the expansion of glass manufacturing in and around Pittsburgh was furthered by local supplies of this fuel. The industry expanded even faster in Ohio and West Virginia where natural gas is also available.[2] Recently the glass region of the Appalachian Plateau has been linked by pipeline with the huge natural gas wells of the Midcontinent Field. It is doubtful if this new source of natural gas will attract any substantial part of the American glass manufacturing capacity as it is far removed from the areas with the highest market potential.

Several cities in western Pennsylvania and New York, southeastern Ohio, and northern West Virginia are either dominated by, or strongly influenced by, glass manufacturing (Washington and Jeanette, Pa.; Corning, N.Y.; Newark and Lancaster, Ohio; and Moundsville and Clarksburg, W. Va.). Besides Pittsburgh among the big cities Toledo has for a long time been known as a glass manufacturing center, producing containers and special glass for automobiles.

In Illinois the small cities of La Salle, Ottawa and Streator have a diversified manufacturing industry based on local supplies of coal, glass sand and clay. Factories of these towns produce brick, tile, and cement besides various types of glass.

## Instruments and Related Products

Manufacturing of instruments and related products[3] is a high labor and low raw material cost industry, which means that transportation costs for raw materials and the finished products play a rather insignificant role for the localization. When this industry, nonetheless, is highly concentrated in the Manufacturing Belt, especially in the older Eastern states, and not in the South with its lower wages, it must be for two main reasons. (1) The higher wages of the Manufacturing Belt reflect a difference in

1. J. R. Smith & M. O. Phillips, *Industrial and Commercial Geography* (New York, 1946).
2. G. E. McLaughlin, *Growth of American Manufacturing Areas* (Pittsburgh, 1938).
3. *Manufacturing of watches and clocks* in 1947 employed 35 thousand people according to the Census of Manufactures (Conn. 10, Ill. 8, N.Y. 6).
*Photographic equipment*, 51 thousand (N.Y. 35, Ill. 7).
*Scientific instruments*, 20 thousand (N.Y. 7, N.J. 6).
*Mechanical measuring instruments*, 53 thousand (N.Y. 12, Pa. 9, Ill. 7, Conn. 5, Mass. 4).
*Ophthalmic goods*, 22 thousand (N.Y. 11, Mass. 5).
*Surgical appliances and supplies*, 22 thousand (Ill. 5, N.Y. 4, N.J. 3, Pa. 3).

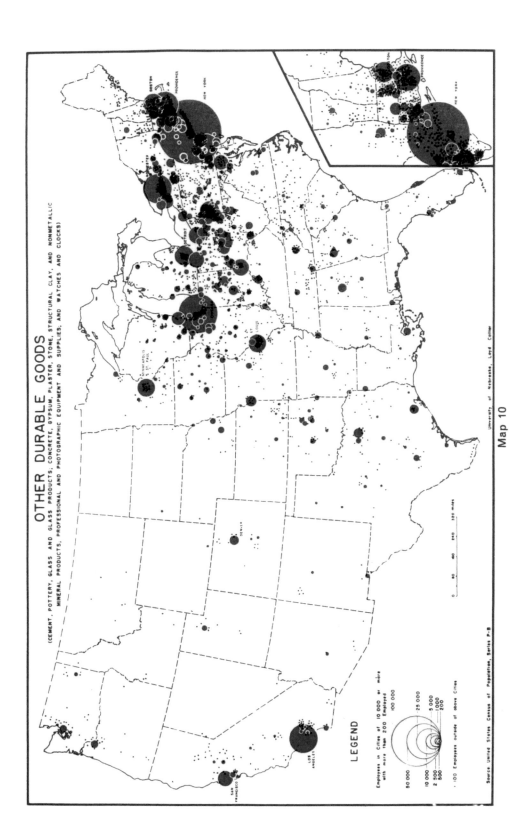

## OTHER DURABLE GOODS

(CEMENT, POTTERY, GLASS AND GLASS PRODUCTS, CONCRETE, GYPSUM, PLASTER, STONE, STRUCTURAL CLAY, AND NONMETALLIC
MINERAL PRODUCTS, PROFESSIONAL AND PHOTOGRAPHIC EQUIPMENT AND SUPPLIES; AND WATCHES AND CLOCKS)

### LEGEND

Employees in Cities of 10 000 or more
with more than 200 Employed

100 000

50 000

25 000

10 000

5 000

2 500

1 000

800

200

· 100 Employees outside of above Cities

0   80   160   240   320 miles

Source  United States Census of Population, Series P-8

University of Nebraska, Loyd Collier

Map 10

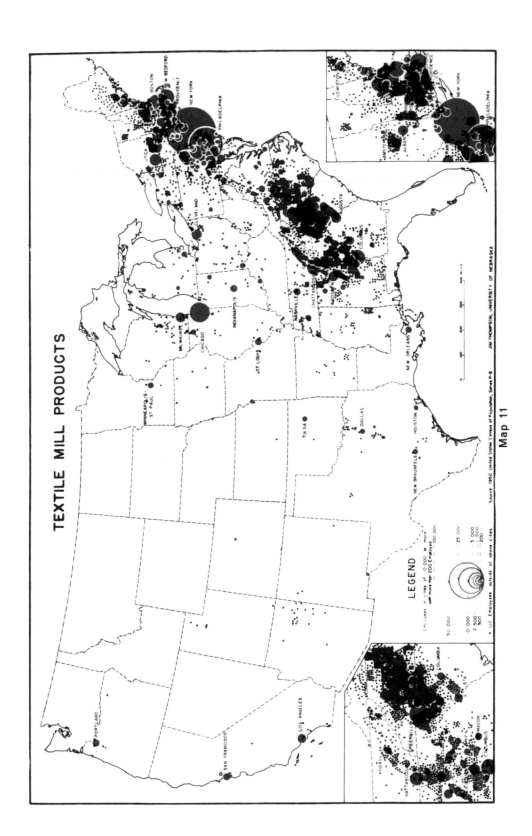

TEXTILE MILL PRODUCTS

Map 11

labor skill, due to decades of experience. (2) Many plants were originally based on inventions and their production is protected by patents. Inventors do not emerge haphazardly in a country. They seem to have a special liking for big cities with their opportunities for technical training and for established manufacturing areas with their technical environment.

The American watch and clock industry can be seen as a justification for the first hypothesis. Connecticut has long had the largest production of watches in the United States. Waterbury, the American brass center, has been known since the end of last century for its cheap watches and clocks. The modern industry was preceded here more than a hundred years by a handicraft industry making wooden clocks. The "migration" of this industry from the oldest center in New England has been west and not south. Elgin, Illinois, in the western part of the "Metal Manufacturing Belt," is one of the other important watch centers. Watch manufacturing in the United States cannot compete on an even basis with the extremely efficient Swiss industry, except for the cheaper varieties of the product.

The photographic equipment industry became concentrated in Rochester, New York, because George Eastman happened to live here. His firm grew from six employees in the early 1880's to about 40,000. Eastman was an amateur photographer with inventive talents and a business genius. He started his firm in 1880 making dry plates. The great expansion came with the introduction of the first practicable roll film in 1884 and the first compact handheld camera, the Kodak, in 1888.

Rochester, at the time when young East-

man started his firm, was one of the largest cities in the highly urbanized belt stretching from New York to Buffalo along the Hudson River and the Erie Canal. This area, one of the major traffic arteries in the United States, has a highly diversified manufacturing industry of the type also found in southern New England, characterized by ingenuity and labor skill. A disproportionately large number of American inventions and discoveries seem to have been made by people living in these two areas.

New York and Chicago are important centers for the manufacturing of different types of instruments and related products. Much of the New Jersey employment in branches of this industry occurs in suburbs of New York City included in the urbanized area of New York.

## Jewelry and Silverware

Providence, Rhode Island, is the leading Anglo-American center for jewelry production. The manufacturing of jewelry and silverware, ranging from cheap trinkets to the best of plate, is as important as the textile industry in the industrial structure of Providence, New England's second largest city, and it has about as long a tradition in the economic history of the city. Other cities in New England known for their silverware are Taunton, Massachusetts, and Wallingford and Meriden, Connecticut. In New York the small city of Oneida in the Erie Canal Belt is a one-sided silverware town.[1]

TEXTILE

Textile manufacturing and the apparel industry, closely associated in the manufacturing process, are both low-wage industries and both employ a large share of women. The two industries differ considerably, however, in their distribution patterns.

---

1. According to the 1947 Census of Manufactures, production of *Jewelry* (*precious metal*) employed 26 thousand people (R.I. 6, Mass. 4, N.Y. 7, N.J. 4).

*Silverware and plated ware,* 22 thousand (Conn. 7, Mass. 3, R.I. 2, N.Y. 6).

Textile manufacturing is the most pro-nounced city forming industry in the United States. No less than 37 of 223 one-sided towns (A-type) are dominated by textile manufacturing. There are also many textile towns of B- and C-type. The apparel indus-try on the other hand, employing almost as many people (Fig. 1) does not dominate any city. It is in its location a "parasite" industry which occurs in a few big cities and in towns dominated by male indust-ries.

Textile manufacturing in the United States is well concentrated in two regions: the Northeast with New England, New Jersey, eastern New York and eastern Pennsylvania; and the Southeast stretching from Virginia through North and South Carolina, Georgia, Alabama and Tennessee (Map 11). The industry includes the three steps of spinning the yarn, weaving or knitting yarn into cloth and finishing the fabric. Some of the end products, about 20 per cent, are finished consumer goods like blankets, knitted underwear, sweaters, ho-siery and squaresewn goods (bed sheets, pillow cases, towels, etc..) Other products are in the form of cloth, which forms the raw material for the apparel industry, about 40 per cent. The remaining 40 per cent go into other industrial uses in the form of cordage, twine, tires, bags, felts, etc.[1]

The industry can be divided into the following five branches: cotton manufactur-ing, woolen and worsted manufactures, rayon and silk production, knitting mills and miscellaneous textile production. As these branches do not coincide in their distribution patterns they will be treated separately here. Mills usually specialize in one fiber: cotton, wool, rayon, silk, etc. Recently, however, with the advent of rayon and nylon, there has been an increasing tendency to mix fibers and to weave the new fibers on the old weaving systems, a practice which tends to obliterate the boundary between the textile branches.[2]

## Cotton

Cotton manufacturing, largest of the textile branches with more than one-third of the total employment, was more than any other manufacturing industry a child of the Industrial Revolution.[3] The invention in England of spinning and weaving machines in the last decades of the eighteenth century made possible the mass production of up-till-then expensive cotton fabrics, and Ely Whitney's invention of the cotton gin in 1793 made cheap American cotton available to the mills in Lancashire. The first cotton mills in the United States were built in New England in the 1790's, but the great expan-sion came with the War of 1812. Protected by a tariff the American cotton industry was able to expand even after the war and to exploit the growing domestic market. The first mills were established on sites where falling water was available as a source of power: at first on small and easily harnessed waterfalls, which abound in the glaciated landscape of New England, later, with in-creasing demand for power, on the bigger falls. On these larger power sites grew textile cities like Lowell and Lawrence, Massachusetts, and Manchester, New Hamp-shire—all three on the Merrimac River. As water remained the dominant source of power for New England's cotton industry during the first three-quarters of the nine-teenth century—the early period of the American Industrial Revolution—the distri-

1. H. E. Michl, *The Textile Industries* (Washington, 1938).
2. E. B. Alderfer and H. E. Michl, *Economics of American Industry* (New York, 1950), p. 445.
3. Even today the textile industry, especially cotton manufacturing, seems to lead industrialization in hitherto non-industrialized countries. Coarse cotton fabrics meet an essential human demand and they can be produced by machines imported from the old cotton-exporting countries and attended by unskilled or semi-skilled laborers.

bution of waterfalls has, through industrial inertia, exercised an important influence on the distribution of manufacturing in the old textile districts of the United States. When the water power at interior points was exhausted and supplementary steam power became necessary, tidewater towns like Fall River and New Bedford in southern Massachusetts expanded and became the leading cotton centers. They could get waterborne coal cheaper than inland towns.[1]

The conspicuous increase in American cotton manufacturing before the Civil War was confined largely to the North, especially to New England. In the Southern states a few cotton mills and iron foundries were the only signs of the early Industrial Revolution. The size of Southern cotton manufactures in 1850 was less than one-seventh of that of New England, measured by the invested capital. The South had turned its interest to agriculture, especially the production of cotton for the rapidly growing world market.

After the Civil War American cotton manufacturing industry experienced a relocation of a magnitude which has hardly been equaled either in the United States or abroad. The comparatively rapid shift of the geographical center of cotton manufacturing from New England to the Piedmont region of the South has become the classic example of an internal industrial migration. The total capacity of the cotton textile industry, measured by the number of spindles, continued to rise until 1925, but the rate of increase was higher in the South than in New England. The latter region reached its peak in 1923 with about 19 million spindles, as compared with 16 million spindles in the South. The decline in New England's cotton manufacturing from the mid-Twenties to the outbreak of World War II was very rapid. In 1939 about four-fifths of the American spinning capacity was located in the South, and the ratio for weaving capacity was roughly the same. After 1939 the change in relative importance between the two areas has been slower.[2] The remaining cotton capacity in New England is mainly specialized in the finer fabrics.

The main attraction of the South for cotton manufacturing was an ample supply of low-wage labor, "poor whites," drawn especially from the overpopulated, soil-eroded hill country.[3] With the expansion of the industry the wage differential between North and South has gradually declined, though some differences in average wages persist. Other factors which contributed to the shift were the absence in the South of restrictive social legislation governing working conditions and the Southern practice of granting free land for the construction of mills and relieving the textile mills of taxes for a period of years. Since the average cotton mill in the South was built later than those in New England, it is likely to be located more in accordance with the modern production situation, it is more rationally designed and it is equipped with new and

1. J. H. Burgy, *The New England Cotton Textile Industry* (Baltimore, 1932).

2. According to the 1947 Census of Manufactures the manufacturing of *cotton broad woven fabrics* employed 337 thousand people (S.C. 86, N.C. 71, Ga. 65, Ala. 35, Mass. 27, Va. 15, Me. 10).
The *cotton yarn mills* employed 92 thousand people (N.C. 54, Ga. 14).

3. In the unimportant textile industry of the South before the Civil War the mills were often owned by planters who used their own slaves. In the expansion after the war Negroes were, however, excluded from work in the textile mills with the exception of some menial tasks, such as sweeping, cleaning and yard labor. The proportion of Negro workers in Southern textile mills and clothing factories declined from 7 per cent in 1920 to 4 per cent in 1940. The explanation seems to be that the origin of the new cotton industry in the South was not a matter of individualistic enterprise alone. Regular "cotton mill campaigns" were organized by "citizens' committees" which often raised funds for the purpose. The entrepreneurs depended on the backing of white citizens and these were influenced by anti-Negro sentiments fed by the Reconstruction period. (Myrdal, *An American Dilemma*, New York, 1944, p. 1111).

Textile Towns

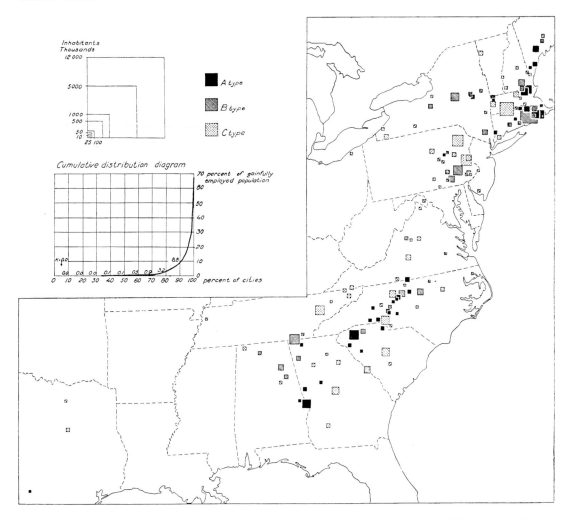

FIG. 12. The textile towns form a string from New England to the southern part of the Appalachian Piedmont.

more efficient machinery. This was an additional competitive advantage of Southern mills in the transitional period.

Contrary to the situation in New England, where cotton mills are concentrated in a few cities, the Southern mills are widely scattered, often located in small towns which have grown up around their single factory, the cotton mill. This is strikingly brought out on Map 11, on which dots almost blacken out the red circles representing cities in the Carolinas, Georgia and Alabama. The dispersed cotton mills of the South seem to be a logical answer to the industry's search for low-wage locations. Wages tend to be lower in small towns than in big urban agglomerations. Among the many towns with more than 10,000 inhabitants dominated by textile manufacturing in the Southeast two cities stand out through their size: Greenville, South Carolina, and Columbus, Georgia.

The relocation of the cotton textile industry had serious economic and social consequences for the old cotton districts in New England, where in many communities cotton manufacturing had been the main industry. The situation was complicated by a general overexpansion of the cotton manufacturing capacity, which during most of the inter-war period made the cotton industry stand out, along with coal mining, as a prime example of a sick industry. In Great Britain an even worse situation was caused in the same period by a decline in exports as a result of industrialization in the importing countries and by increased competition from low-wage nations, mainly Japan. In the United States neither imports nor exports bulk large in comparison with the output for the domestic market. The depressed conditions of American cotton manufacturing were generated within the industry itself. Cotton manufacturing ranks with farming and bituminous coal mining as one of the most competitive industries in the American economy. All three had a considerable overcapacity even during the generally prosperous years of the 1920's.

## Wool

The wool textile industry in the United States, as well as in Europe, is older than cotton manufacturing. Before the Industrial Revolution cotton was in the luxury class with silk, whereas wool had been for centuries the generally used textile fiber. Until far into the nineteenth century most of the work with spinning and weaving wool was carried on in the household and thus was an integrated part of the farm family's production. From the earliest times in the colonies there were also, however, people specialized in certain operations in the making of wool fabrics, especially fulling but even weaving. These men would correspond to the gainfully employed population in wool manufacturing if the present standards of collecting employment data were applied to the pre-industrial period.

The first wool factories in the United States were built in the last years of the eighteenth century in New England. Protected by a high tariff wall the industry expanded rapidly during several decades preceding the Civil War. As the skill in wool production was widely diffused, mills were established in most of the states and territories, but about four-fifths of them were located in New England, New York, New Jersey and Pennsylvania by 1870, when the maximum number of plants was recorded.[1] This general distribution pattern has remained substantially unchanged up to the present. New England's woolen and worsted mills employ almost two-thirds of the American total. Lawrence, Massachusetts, is a leading American wool manufacturing center. New York, New Jersey and Pennsylvania have about twice as many employees as the Southern states[2] but there are indications of a shift to the South even of the woolen and worsted industry. Very few mills have been built since World War I, but almost all of them have been erected in the South. The technological developments in the worsted branch of the industry have been so extensive in recent years that owners of the more obsolete plants face the necessity of building new plants. It is likely that these mills will be built in the South,[3] but the industry will hardly experience a regional migration of the magnitude encountered in cotton manufacturing.

1. Alderfer and Michl, *Economics of American Industry* (New York, 1950), pp. 385 ff.
2. According to the 1947 Census of Manufactures the woolen and worsted industry employed 180 thousand people. Of the 132 thousand in mills making fabrics, 82 thousand lived in New England (Mass. 40, R.I. 13), 28 thousand in the Middle Atlantic states (N.Y., N.J. and Pa.) and 14 thousand in the South.
3. Alderfer and Michl, *op. cit.*, p. 402.

## *Rayon and Silk*

The rayon and silk industry in the United States has undergone a profound change during the last decades. Rayon production has followed a steeply rising curve, whereas silk manufacturing, reaching its peak development in the Twenties, has disappeared as a separate industry, at least in the statistics.

American silk manufacturing grew rapidly from the 1860's as a result of several national and international events. The elimination in 1860 of tariff protection for silk products in Great Britain virtually destroyed the British industry, and the unemployed throwsters and weavers emigrated to the United States. At about the same time the United States imposed a high tariff on silk products and eliminated the duties on raw silk, which made the raw material, imported from the Orient (mainly Japan), cheaper for the domestic manufacturers. Also at about the same time the power loom was introduced to silk manufacturing in Switzerland and it was quickly adopted by the American industry. The rapid growth of silk manufacturing in the United States was continued into the 1920's, when over half of the world production of silk fabrics was produced and consumed in this country. In 1919 the American silk factories employed over 126,000 people.

During the early part of the expansion period the industry was largely concentrated in New Jersey, chiefly in Paterson, which became known as the "silk city." Paterson is located on the outskirts of the present urbanized area of New York. The proximity to New York, the American style and apparel center, seems to have been the most important locational advantage of Paterson. Silk fabrics have always been the luxury product of the textile industry and they are exposed to rapid style changes. As a result, the industry is characterized by small plants, smaller than in the wool industry, which also produces style fabrics. For mills in the final stage of silk production proximity to the market is essential.

Most of the throwing ("spinning") mills were, from the 1880's, built in eastern Pennsylvania in the anthracite cities, like Scranton and Wilkes-Barre, and the cement centers, Allentown and Easton. In these cities, dominated by male industries, was found an adequate supply of cheap female labor. The shift of silk throwing from Paterson with its high wages was later extended to weaving. Even the silk district of eastern Pennsylvania is close to New York.

The rapid decline of the silk industry after the 1920's was caused mainly by the rise of rayon or artificial silk as a weaving material.[1] Commercial production of artificial silk was begun in the United States in 1911. As early as 1926 consumption of rayon—a word adopted by the industry in 1924 as a substitute for the prejudicial "artificial silk"—exceeded that of silk, and in the late Thirties even wool was surpassed. The production of synthetic fibers—rayon, nylon, dacron, orlon, etc.—is a branch of the chemical industry, whereas the weaving of these fibers is a textile industry.

Rayon weaving is widespread in the American textile districts.[2] Both the silk and the cotton industries could easily adjust their equipment to the new fiber. The fine-goods cotton mills of Massachusetts were the first to shift to rayon weaving, and later

1. The 1947 Census of Manufactures reports 1,800 employees in silk yarn mills. The weaving mills are not reported separately but grouped with the rayon mills. The production of silk cloth in 1947 was 25 million square yards, which should be compared with 88 million in 1899 and 310 million in 1919.
2. In 1947 *Rayon and related broad woven fabric mills* employed 98 thousand people according to the Census of Manufactures: Northeast 27, Middle Atlantic states (N.Y., N.J. and Pa.) 24, South 47. Leading states: S.C. 20, Pa. 16, N.C. 16, Mass. 11. Among fabrics related with rayon, nylon and silk were the most important.

Southern mills followed. Also built were special rayon mills, but few of these were located in the old silk area. Some silk mills even changed to rayon, but their equipment was not as efficient as that of new rayon mills or converted cotton mills. The silk plants were located in high-wage areas—an additional disadvantage especially in comparison with Southern mills.

Although rayon first appeared as a substitute for silk, the market for which it successfully conquered by the late Thirties, the new material also established competition in many fields with cotton and wool fabrics, especially after the introduction of rayon staple. The end of this fight cannot yet be seen. During the Thirties rayon staple had its greatest development in the autarkical countries of Germany, Italy and Japan, which together produced almost 90 per cent of the 1939 world production. The rapid expansion of this fiber in the United States in recent years, in keen competition with wool and cotton, has shown that rayon staple is not merely an *ersatz* product. Production has increased even more rapidly than that of filament rayon yarn. Rayon staple has increased the versatility of rayon, which now can be used either to blend with the natural fibers or to simulate cotton, linen, worsted and silk fabrics. The market for rayon thus has been greatly broadened.

*Knitting*

In number of employees the knitting industry is second only to cotton manufacturing among the textile branches.[1] About 60 per

cent of the employees work in hosiery mills; 10 to 20 per cent are employed in knit underwear and in knit outerwear mills. The knitting mills have a wider geographic distribution than the other textile branches. Several Midwestern states, otherwise almost devoid of textile industries, rank relatively high in this branch.

North Carolina has more people employed in hosiery production than any other state. Mills are scattered throughout the many small towns of the Piedmont, but the chief concentration is in or near Burlington and High Point. Tennessee and Georgia are also among the leading hosiery states. The South has more than half of the hosiery mill employees of the United States, with a great dominance in the seamless hosiery field. This is still another indication of the importance of wage differentials as a locational factor in the branches of textile and apparel manufacturing which are not tied to the fashion centers. The full-fashioned hosiery mills, chiefly women's hosiery, using nylon as their most important raw material, are more evenly distributed between the Northeast and the South, with Pennsylvania as the leading state. Philadelphia, in the old silk region, leads in the production of women's hosiery.

The knit outerwear branch is strongly concentrated in the Middle Atlantic states, especially New York. Utica in the Mohawk Valley is known as a knitting center. In underwear production no state dominates.

*Miscellaneous*

Under this title the following branches are grouped: the carpet and rug industry, production of hats other than millinery and cloth hats, and miscellaneous textile goods production.[2] They cannot be treated as one group from a locational point of view and only a few remarks will be made here.

The carpet and rug production in the

---

1. *Knitting mills* in 1947 employed 231 thousand people according to the Census of Manufactures: *Full-fashioned hosiery mills*, 70 thousand (Pa. 25, N.C. 16, Tenn. 4); *Seamless hosiery mills*, 65 thousand (N.C. 27, Tenn. 10, Pa. 7, Ga. 6); *Knit outerwear mills*, 35 thousand (N.Y. 15, Pa. 6); *Knit underwear mills*, 41 thousand (Pa. 10); *Other knitting mills*, 20 thousand.

2. The *carpet and rug industry* employs 57,000 people, *the hat industry* 21,000 and *the miscellaneous textile industry* 60,000.

United States is strongly concentrated in the Middle Atlantic States, close to the other wool-using industries. Amsterdam in the Mohawk Valley and Philadelphia are the leading centers. The production of fur-felt hats is a specialty of Danbury, Connecticut.

APPAREL

The American apparel industry is highly concentrated in the urbanized area of New York, which has 34.7 per cent of all gainfully employed in the industry or 369,000 out of 1,063,000 (Fig. 1). Apparel manufacturing is by far the largest employer among the New York manufacturing industries, giving work to 7.4 per cent of the gainfully employed population of the city. This makes New York an apparel center of C-type. The only other manufacturing industry with a similar geographic concentration is the automotive industry with 296,000 people, or 34.1 per cent of the national employment in Detroit.

The transformation of garment making from a craft or household status to a manufacturing industry began in the United States on a small scale about 1825. The first ready-to-wear clothing found a market among sailors. New York, Boston and New Bedford[1] became the principal centers for this trade since they were leading ports. Designing and cutting were done in shops and most of the sewing was carried on in the homes. Cheap clothing for the Negro slaves of the South and equipment for California gold miners were important bases for the early expansion of the industry.

A milestone in the history of apparel manufacturing was Elias Howe's construction in the 1840's of a practical sewing machine, later improved by Singer and others. The industry, then, was mechanized when the boom of the Civil War started. During the war Army authorities collected physical measurements of millions of soldiers, which furnished the statistics necessary for standardized clothing sizes. With the improved fitting qualities a permanent and expanding market was established for ready-made garments.[2]

Labor for the industry was provided by women and children and by the steady influx of foreigners, first mainly the Irish, after 1876 Russian Jews, and later also Italians. The immigrants, often arriving without funds and knowledge of English, kept the supply of labor in the main port of entry ahead of demand, with resulting low wages. In this industry developed what has been termed the "sweatshop system," under which operatives worked for very low wages in their own homes. Through public interference the industry was forced out of the overcrowded and unhealthy tenement houses into small factories, and from about 1910 the conditions of the garment workers were improved, partly because their unions gained in strength after decades of failures, partly because immigration soon afterward was restricted.

Apparel manufacturing, from a locational point of view, has two main characteristics. It is a big-city industry and it is a low-wage industry. These two seemingly incompatible features can be derived from the dominant influence of style on the industry. Big cities are style-setting centers. As the industry is so dominantly influenced by rapidly shifting fashions, it has not been mechanized in a way comparable with other manufacturing industries. Labor cost is thus a very important cost item. As the biggest item of cost, the fabric, tends to be the same for all

---

1. New Bedford was the largest whaling center in the United States at a time when whaling was a major American industry. S. L. Wolfbein, *The Decline of a Cotton Textile City. A Study of New Bedford* (New York, 1944), p. 7.
2. Alderfer and Michl, *Economics of American Industry* (New York, 1950) pp. 447 ff.

firms, except for a slight advantage to very large purchasers, labor cost is the most important competitive factor. Low wage districts—cities with predominant male industries and the Southern states—exert a strong pull, especially on plants in the low-price lines, which are not so strongly tied to the fashion centers.

On Map 12 the attraction of low-wage areas is evidenced by a large number of dots throughout the Southern states, some in the iron ore mining districts on Lake Superior, and by large circles for the anthracite cities of Scranton and Wilkes-Barre, just to mention a few examples. The dispersing effect of the industry's trend towards low-wage areas is rather well counterbalanced by the concentrating force exerted by fashion centers—big cities, especially New York. There has, however, long been a substantial migration of the apparel industry to smaller cities of various industrial structures but located not very far from New York, usually within overnight train or truck distance, in New Jersey, eastern Pennsylvania, eastern New York, Connecticut and southern Massachusetts. The result of this development is clearly brought out on Map 12 and Fig. 14.

For an understanding of New York's dominant position within this industry it is necessary to study in a little more detail the structure of New York's garment manu-

facturing. The city produces about 40 per cent of men's clothing and no less than 70 per cent of women's apparel, a situation which seems to be a reflection of the greater importance of fashion for women's garments. This is further reflected by a smaller average size of plant in the women's clothing industry than in the men's. In the dress industry, the most important branch, the average inside shop in New York employs about 40 workers and the average contractor about 25 workers.[1]

The New York clothing plants are highly concentrated in the small area of Manhattan bounded by Twenty-fifth Street on the south, Forty-second Street on the north, Eighth Avenue on the west and Fifth Avenue on the east. Here is found a geographic concentration greater than in any other manufacturing. About three-fourths of the 369,000 New York garment employees work in an area about 200 acres in size. But, unlike the case in most other manufacturing industries, the geographic concentration is brought about by a clustering of thousands of small plants. The problems arising from the rapid changes of style are more easily mastered by small units than by the huge firm with its top-heavy, bureaucratic organization. The individual producer is highly specialized, not only to a particular type of garment but often also to a specific price class within this type, and therefore he is dependent upon other producers and auxiliary services, which are available close at hand in the New York clothing district. The concentration of thousands of plants in a small area allows every manufacturer to be in touch with and get "ideas" from the others. For buyers coming from all over the country it is convenient not only to have such a predominant portion of the total production concentrated in one city but to have most of the showrooms located in a few blocks.[2]

1. The jobber-contractor system developed in the nineteenth century to recruit and utilize immigrant labor. It is still most heavily concentrated in the women's wear branch of the clothing industry. About 80 per cent of the dresses produced in New York are made by contractors. In this system the clothing jobber buys the cloth, selects the style and cuts the garment, which is then handed over to the contractor to be sewn and finished into the final garment at a set price. Alderfer and Michl, *Economics of American Industry* (New York, 1950), pp. 447 ff.

2. During the nineteenth century large quantities of fine clothing were imported from Europe. Merchants from all parts of the United States went to the landing port, New York, which became established as the leading wholesale center. This historic development goes far to explain why New York and not any of the other big cities became the American apparel center.

Apparel Towns

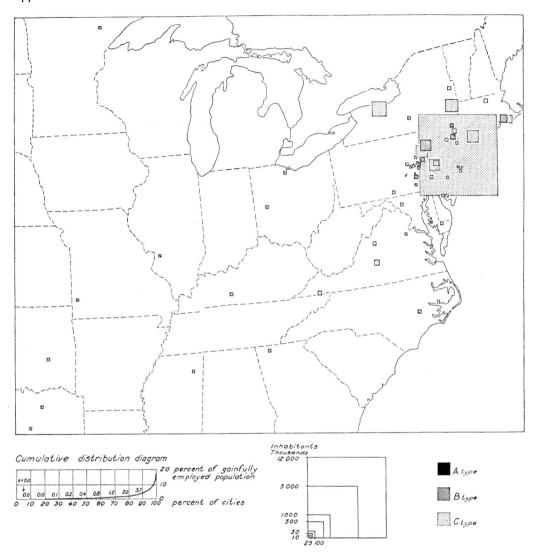

FIG. 13. New York is the dominant American apparel city. There are a large number of smaller cities near New York, in New Jersey, eastern Pennsylvania, southern New York State, and southern New England, in which the apparel industry is important and to which many New York firms have migrated.

Outside of New York some of the other big cities also have a large clothing industry. Philadelphia, Los Angeles, Chicago, Boston, St. Louis, Cleveland and Baltimore are important in one or more of the clothing branches, usually in men's clothing, in which New York's dominance is not so great.

An interesting development during the last decades has been the emergence of Los Angeles as a fashion center with a large apparel industry, especially in women's clothing, which must be seen against the background of Los Angeles position in the movie industry. Another city which is

# APPAREL INDUSTRY

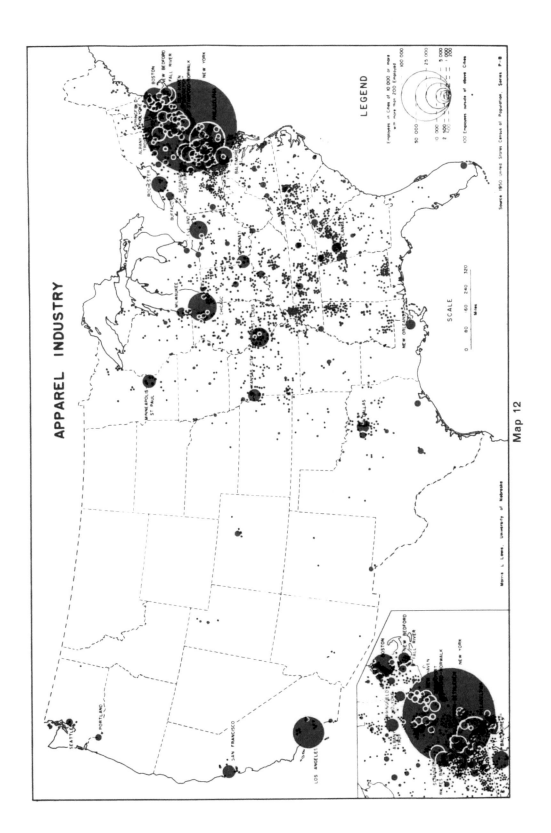

## LEGEND

Employees in Cities of 10,000 or more
with more than 200 Employed

100,000

50,000
25,000
10,000
5,000
2,500
1,000
500
200

100 Employees outside of above Cities

Source 1950 United States Census of Population, Series P-8

SCALE

0    80    160    240    320

Miles

Morris L. Lewis, University of Nebraska

Map 12

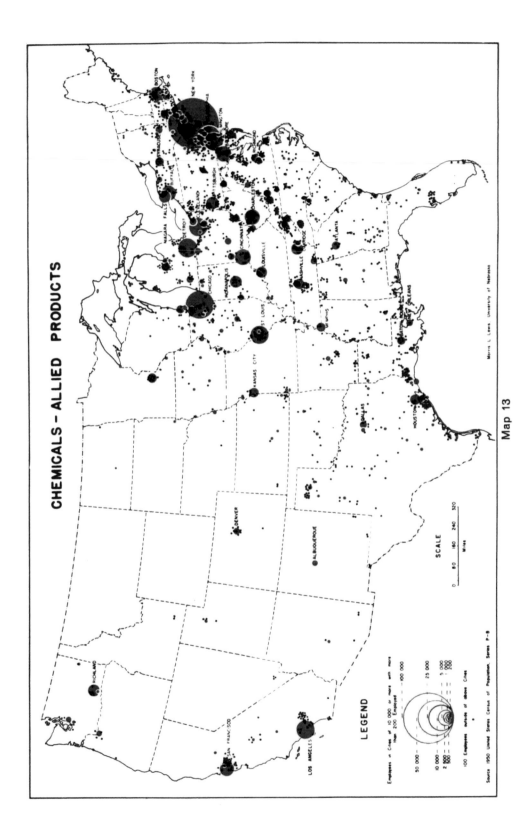

CHEMICALS – ALLIED PRODUCTS

LEGEND

Employees in Cities of 10 000 or more with more
than 200 Employed

100 000

50 000

25 000

10 000
5 000

2 000
500

100 Employees outside of above Cities

SCALE

0    80    160    240    320

Miles

Source  1950  United  States  Census  of  Population,  Series  P-8

Morris L. Lewis, University of Nebraska

Map 13

rapidly growing in importance as a fashion and apparel manufacturing center is Dallas, Texas.

Among the cities of intermediate size Rochester, N.Y., is known for high quality men's clothing. Troy, New York, through one of many historic accidents which have greatly influenced the distribution pattern of apparel manufacturing, as well as that of other industries, became the national center for the production of collars and cuffs from about 1850. The city still ranks high in the manufacture of shirts and collars.

CHEMICALS

The chemical industry is expanding at a faster rate than any of the major manufacturing groups in the United States. The most spectacular development has been in the new field of atomic energy production, but the branches of longer standing are also dynamic. The biggest chemical enterprise in the United States, du Pont, reports that about half of its output represents products which were unknown or in their commercial infancy in the early 1930's. Du Pont spends about three per cent of total sales value on research, a considerable amount of which goes to fundamental research.[1] The American chemical industry for a long time could build on scientific knowledge accumulated in the older European industry, especially that of Germany. But during the last three decades the big American companies have spent increasing amounts of money on fundamental research. The most revolutionary product of such research in the United States, before the atom and hydrogene bombs, was nylon, which was ready for commercial production in 1938 after more than ten years of work by a du Pont group under Carothers.

Chemical plants are widely scattered over the Manufacturing Belt and the entire South with the heaviest concentration in the area from New York to Baltimore on the Atlantic Seaboard. From a locational point of view the industry can be divided into light and heavy chemical manufacturing. Heavy or industrial chemicals (sulphuric acid, soda ash, caustic soda, chlorine and ammonia), selling at low prices relative to weight and bulk, cannot carry the costs of long transportation. They are manufactured near sources of raw materials or in port cities in the market areas. Light chemical industries, exemplified by pharmaceutical preparations, make expensive products relative to weight and can locate themselves with regard to the national market.

Large-scale enterprises are a characteristic of the chemical industry. The biggest and most succesful of the corporations operate in a number of fields. The consolidation has, however, not gone so far in the United States as in Germany (IG Farben) or Great Britain (Imperial Chemical). The structure of the industry as well as its youth, suggest a distribution pattern which is more rational with regard to the present situation than that of older industries like primary iron and steel, which were localized over a long period of time. Radical changes in transportation and fuel economy, etc. may have made old centers of such industries obsolete, even if they remain on the map as a result of industrial inertia.

New York is the largest chemical center on the American continent. Most branches of chemical manufacturing are represented here. The important production of heavy chemicals is largely located on the New Jersey side of the Hudson River, where sites are available for the plants, many of which are very large. The port location results in relatively low assembly costs for raw materials. Most of the finished products

---

1. *Du Pont, the Autobiography of an American Enterprise* (New York, 1952), pp. 120ff.

are used in industrial processes. The New York area, the biggest manufacturing center in the world, offers a huge market. Production of finer chemicals, particularly pharmaceuticals, is concentrated in Brooklyn.[1]

Philadelphia, the second largest port city in the United States, has advantages for chemical manufacturing similar to those of New York. Several big chemical plants are located on both sides of the Delaware River. One of its tributaries, the Brandywine Creek, holds a prominent position in the history of American chemical manufacturing. It was here, on the outskirts of Wilmington, that the French emigrant family du Pont in 1802 started a powder mill, using water power to turn their machinery. The Brandywine gorge was one of the earliest districts of concentrated manufacturing in the United States.[2] When du Pont established his mill, the many small falls and rapids of the creek supplied power to a great number of mills. Large chemical plants and the headquarters of du Pont, as well as its main experimental station, now make Wilmington a chemical city of B-type.

Baltimore has long been known as a very important fertilizer center. Waterborne raw materials, phosphate from Florida and pyrites from Cuba, are unloaded directly at the wharves of the plants.[3] The finished products are shipped mainly to the Southeast and to the Middle Atlantic states, the two areas with the heaviest consumption of fertilizers in the United States. Practically all ports between Maine and Texas have fertilizer plants. Production is rapidly increasing in the Middle West and California.

A conspicuous string of chemical plants stretches in the Appalachians from northern Virginia into Tennessee. Eastern Virginia also has a concentration of chemical establishments in the Richmond-Hopewell area. The South, and especially Virginia, is the center of the rayon industry and of the new synthetic fiber production (nylon, orlon, dacron), but other types of chemical plants are also located here. A copious supply of soft water, important in several of the chemical industries, a large labor reserve, a strategic location with regard to markets and ready access to raw materials have attracted the chemical industry to Virginia. The greatest attraction of Tennessee seems to have been cheap hydro-electric power from the TVA. The huge nuclear energy plants at Oak Ridge, which have created a new town of 30,000 inhabitants, were located in the TVA area because of the tremendous amounts of electrical energy needed in atomic research.[4]

The Great Kanawha Valley in West Virginia, centered on Charleston, is known locally as the "Chemical Valley." Large plants are closely spaced along this tributary of the Ohio River. The chemical industry is the outstanding manufacturing activity of the valley, to which it was attracted during World War I by the natural advantages of coal, natural gas, petroleum, salt, hydro-electric power and water.[5] Production is diversified with an emphasis on heavy chemicals. The Ohio River also has some important chemical centers along its shores: Cincinnati, Louisville and Parkersburg, West Virginia.

1. The chemical industry is represented in the 1947 Manufacturing Census by a number of small branches. Only the largest of these can here be specified as to most important states. *Synthetic fibers*, 72 thousand employees (Va. 21, Tenn. 14, Pa. 8). *Organic chemicals, n.e.c.*, 85 thousand (N.J. 20, W.Va. 13, N.Y. 10, Tex. 8, Mich. 8). *Pharmaceutical preparations*, 66 thousand (N.Y. 12, Ind. 9, N.J. 8, Mich. 8, Ill. 7). *Paints and varnishes*, 53 thousand (Ill. 7, N.J. 7, N.Y. 6, Ohio 6).
2. *Industrial Cities Excursion, Guidebook.* XVIIth International Geographical Congress (1952), p. 26.
3. Pearle Blood, "Factors in the Economic Development of Baltimore, Maryland." *Econ. Geogr.* (1937).
4. *Southeastern Excursion, Guidebook.* XVIIth International Geographical Congress (1952) p. 109.
5. L. M. Davis, "Economic Development of the Great Kanawha Valley," *Economic Geography* (1946).

## Chemical Manufacturing Towns

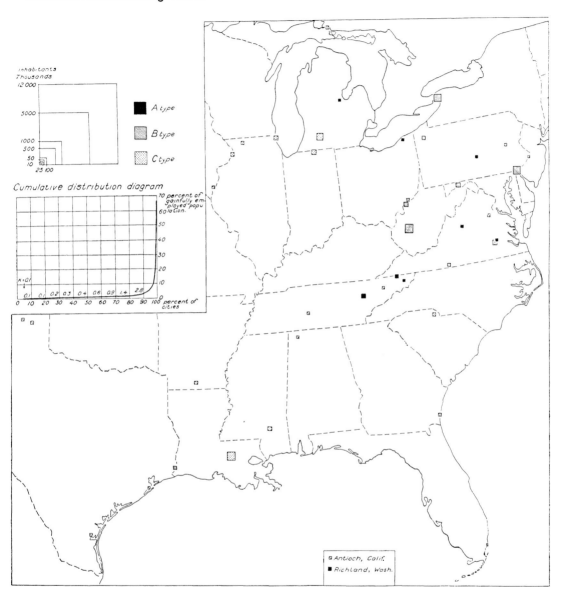

FIG. 14. Cities in which the manufacturing of chemicals is an important or dominant city forming industry are scattered over a wide area in eastern and southern United States. Important chemical towns are Wilmington, Delaware, Charleston, West Virginia, Niagara Falls, New York and Baton Rouge, Louisiana. The "atom cities" of Oak Ridge, Tennessee, and Richland, Washington, should also be mentioned.

A relatively new chemical district stretches along the southern shore of Lake Erie with Cleveland as the main center. The advantages for heavy chemical manufacturing are here the same as for the primary iron and steel industry: an excellent market location and low assembly costs for raw materials. The chemical industry of Detroit has the special advantage of large deposits of salt underlying the city. In east-central Michigan the small city of Midland is a one-sided chemical town with, among other things, the American pioneer plant for production of magnesium, which is made by an eletrolytic process separating the metal from chlorine. The Midland industry is based on local brine. The hydroelectric power of Niagara Falls has attracted many large electrochemical plants making solvents, bleaches and other ingredients used in industrial processes.[1]

Chicago, St Louis, and Indianapolis are important chemical centers in the Middle West.

An interesting and recent development in the chemical field has been the establishment of a large number of plants on the Gulf Coast (especially along the Intracoastal Canal). These are of two types. Inorganic chemical industries have been attracted by the abundance of sulphur, salt, lime, and cheap fuel. Their bulky and cheap products can be shipped by water. The hydrocarbon chemical industry, based upon natural gas and petroleum, made very rapid progress during the war with the increased demand for aviation gasoline, which revolutionized petroleum refining. In the new field of petro-chemical industry the United States holds a position similar to Germany's in the coal-tar industry for several decades after about 1870. The petro-chemical industry produces an amazing number of synthetic products and as yet it seems to be only in its beginning stage.[2]

In the west, Los Angeles and San Francisco are important chemical centers. Richland, Washington, is an extremely one-sided chemical town of the same type as Oak Ridge, Tennessee. The agglomeration was built at the end of the war by the government, eight miles south of the security area of the Hanford atomic works, to house the employees of the Atomic Energy Commission who were moved here from the temporary city of Hanford. Work on the Hanford project was started in 1943, and at its peak some 50,000 men were employed. It was located here because of available power from the Columbia River, abundant water supply, and protective isolation. In peacetime the project is operated by General Electric.[3]

OTHER NONDURABLE GOODS

*Paper and Allied Products*

The modern pulp and paper industry[4] using wood (predominantly softwood) as raw material was largely confined to New England and the state of New York during its earliest development. Here an adequate supply of spruce pulpwood was available, and this district was also favorably located with respect to markets. In the Nineties the industry expanded in Wisconsin, Michigan and Minnesota. Until after the First World War these regions dominated the pulp and paper industry in the United States. Further expansion then became difficult as the industry had reached the limit of available spruce pulpwood supplies.[5] Imported Canadian pulpwood and wood pulp, as well as wood pulp from Scandinavia and Finland, increased in importance. The removal of

1. *Industrial Cities Excursion, Guidebook.* XVIIth International Geographical Congress (1952), p. 109.
2. Elmer H. Johnson, *The Industrial Potential of Texas.* Bureau of Business Research, University of Texas (Austin, 1944).
3. Otis W. Freeman and Howard H. Martin, *The Pacific Northwest* (New York, 1954).
4. *The paper and allied products industry* in 1947 employed 450 thousand people, according to the Census of Manufactures. The largest branches were:
   *Pulp mills,* 50 thousand (Wash. 6, Me. 5, Wis. 5, Va. 4, La. 4).
   *Paper and board,* 148 thousand (N.Y. 15, Wis. 13, Ohio 12, Mich. 11, Pa. 11, Me. 9).
   *Paperboard boxes,* 110 thousand (N.Y. 20, Ill. 11, Pa. 11, Ohio 9, Mass. 8, N.J. 7).
5. C. W. Boyce, "The Pulp and Paper Industry." In Glover and Cornell, *The Development of American Industries* (New York, 1946).

duties on newsprint just before the First World War was followed by a "migration" of the American newsprint industry to Canada. The Canadian production of newsprint in 1913 was about one-third of that of the United States; in 1940 it was more than three times as large.[1]

In the inter-war period, the Pacific Coast states Washington and Oregon had a spectacular increase in the production of wood pulp, especially sulphite pulp. This area is unique in that it ships a sizable portion of its pulp to other sections of the United States.[2]

The most rapid increase in wood pulp production has, however, occurred in the South, where now about half of the national output originates. The South produces chiefly sulphate pulp. The great strides made in the Southern industry since about 1930 have been connected with the name of C. H. Herty, whose researches showed that the high resin content of the Southern pine did not preclude its use for pulping purposes.[3] Most of the pulp is converted into wrapping paper and paperboards in large, integrated mills. Since 1940 at least two huge newsprint mills have also been completed in the South. Some authors believe that this is just a beginning. Production and consumption of pulp is balanced, with only a relatively small portion moving out of the area. The Southern industry is located chiefly along the Atlantic Coast from northern Florida to Virginia with Savannah

and Jacksonville as important centers, along the Gulf Coast east of the Mississippi and in northern Louisiana and southern Arkansas.[4]

The old pulp and paper districts in New England, the Middle Atlantic states, and the Great Lakes states have also increased production during the last decades, when their relative portion of the national output declined. In Massachusetts, where Holyoke is the leading American center for the production of fine writing paper, Connecticut, New York, Pennsylvania, Michigan, and Ohio, the paper and board mills are predominantly non-integrated. The advantages of integration between pulp and paper mills are greatest in the production of newsprint and of most wrapping paper and board, which are made in large quantities to uniform specifications and from only one or two kinds of pulp.[5] The integrated mills therefore are raw material oriented and are largely located in the peripheral regions: the South, the Pacific Northwest, north-central Wisconsin, and the highlands of Maine, New Hampshire and New York. For many products in which different types and grades of pulp are used, a location close to the market is more essential. The Manufacturing Belt with a relatively small pulp production, is thus the leading paper producing region in the United States, and it is to this area that the large imports of pulp are destined. The ports of the Atlantic seaboard and towns on the Great Lakes are favorably located for import of wood pulp from Canada and northern Europe.

## Petroleum Refineries

Petroleum refineries[6] are located with regard to (a) the center of crude oil production, (b) markets, and (c) means of transportation. Location near the market is mainly based on the fact that crude oil can be transported cheaper than the refined product, whereas

1. Sven A. Anderson, "Trends in the Pulp and Paper Industry," *Econ. Geogr.* (1942).
2. John A. Guthrie, *The Economics of Pulp and Paper* (Pullman, Wash. 1950).
3. Anderson, *op. cit.*
4. Th. Streyffert, "Förenta Staternas massa- och pappersindustri med särskild hänsyn till dess råvaru-försörjning." *Kungl. Skogshögskolans skrifter* (1952).
5. Guthrie, *op. cit.*
6. *Petroleum refining* in 1947 employed 146 thousand people, according to the Census of Manufactures (Tex. 36, Calif. 19, Pa. 15, Ind. 12, La. 11, N.J. 11, Ill. 9, Okla. 7, Kans. 4, Ohio 4).

plants located near the oil fields have the advantage that the refined products can be sent in different directions with a minimum of backhauling. As the industry is dominated by big companies—over nine-tenths of the wage earners are in concerns with two or more plants[1]—it may be assumed that decisions concerning the location of new plants are based on scientific investigations.

The American refining capacity is highly concentrated in a few small areas. The port cities of the western Gulf Coast, chiefly Houston, Baytown, Texas City, Port Arthur, and Beaumont in Texas, and Baton Rouge and Lake Charles in Louisiana, have about one-third of the national employment in the refining industry. These plants, located on tidewater, get crude oil through pipelines from the Gulf Coast and midcontinent oil fields. The refined products may be shipped in different directions by tankers: gasoline to the Northeast, diesel oil to Europe or South America, kerosene to China, etc. Several of the inland cities in Texas, Oklahoma and Louisiana have refineries, but with a more moderate capacity. Tulsa, Oklahoma, is the largest of these inland centers.

In the Northeast, the largest concentrations of petroleum refineries occur on the Delaware River in southeastern Pennsylvania and in the urbanized area of New York, mainly on the New Jersey side of the Hudson River. Philadelphia and New York have the advantages of a large local market, as well as a big hinterland. They can absorb most of the refinery products and can get crude oil cheaply by tanker from the Gulf ports. During the war German submarines threatened to break this American life line of tankers carrying crude oil and refinery products from the Gulf Coast to the big refineries and markets of the Northeast, one of the most important sea routes in the world. In this emergency two pipe lines of unheard-of dimensions were built by the government, Big Inch from Longview, Texas, to Phoenixville, Pennsylvania, with branch lines to the refineries of Philadelphia and New York, and Little Big Inch with a similar stretch. After the war these giant pipe lines were turned over to natural gas companies. Water transportation still is the cheapest way of transporting petroleum products and almost all the oil to the large Northeastern market now moves by tanker.[2]

The third big refining and storage center in the Manufacturing Belt is the Whiting-South Chicago district in the urbanized area of Chicago. Whiting is located in Indiana. This district receives the oil by pipe line from the Midcontinent field.

In California, a leading oil refining state, Los Angeles and San Francisco have most of the refining capacity. Through a happy coincidence the large refineries of Los Angeles are both raw material and market oriented. With its rapid increase in population California now consumes a much larger share of its oil production than formerly, and this is reflected in a smaller trade through the ports of Los Angeles and San Francisco, both of which have petroleum products as the leading outmoving commodity.

### Tobacco

Tobacco manufacturing[3] should, from the locational point of view, be divided into

1. Willard L. Thorp and others, The Structure of Industry, App. A, pp. 211–225, Temporary National Economic Committee, Monograph 27, (Washington, 1941).

2. Erich W. Zimmermann, *World Resources and Industries.* (New York, 1951).

3. Total employment in the *Tobacco manufactures* was 112 thousand people in 1947, according to the Census of Manufactures.

*Cigarette manufacturing*, 28 thousand (N.C. 14, Va. 8, Ky. 4).

*Tobacco stemming and redrying*, 26 thousand (N.C. 14, Va. 6, Ky. 4).

*Cigar manufacturing*, 47 thousand (Pa. 18, Fla. 9, N.J. 4).

*Chewing and smoking tobacco*, 11 thousand.

FIG. 15. Petroleum refining towns are located on or near the Gulf Coast, tobacco towns in North Carolina and Virginia, paper and pulp towns in northern New England, Wisconsin, the South and the Pacific Northwest. The dominant rubber town is Akron, Ohio. Several shoe and leather towns are located in New England with Boston as the largest center. Binghamton, New York, is a large shoe and leather city.

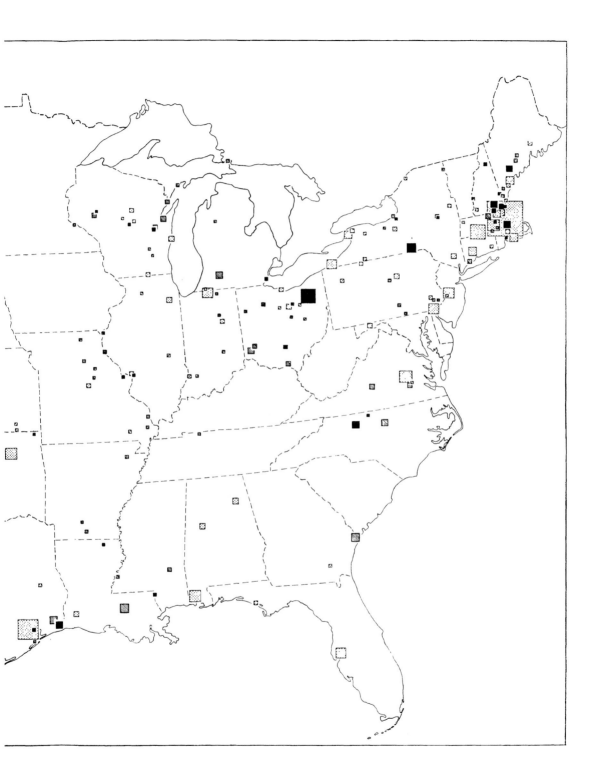

several branches: cigarettes, cigars, smoking tobacco, chewing tobacco, and snuff. Of these, cigarette and cigar manufacturing are the only important ones. They employ about the same number of people, but the cigarette industry is by far the largest, measured by tobacco consumption or value added. It is much more mechanized than cigar production.

The cigarette industry is concentrated in the Piedmont region, with Kentucky (Louisville) as a secondary center of production. Cigarettes became popular after the Civil War with manufacturing firmly rooted in the raw material producing areas—the bright-tobacco growing region of North Carolina and Virginia. The economy of mechanization and large-scale production led to a monopoly as early as the 1890's, but this was dissolved in 1911 through an order from the Supreme Court.[1] Since then the industry has been dominated by the "Big Three," American Tobacco, Reynolds and and Liggett Myer. The factories are located in Winston-Salem, Durham, and Reidsville, North Carolina, and in Richmond and Petersburg, Virginia.

Cigar manufacture depended exclusively on hand labor to the end of the First World War but, in the inter-war period, mechanization spread rapidly and machine-made cigars now have driven the handmade product out of the market except for the higher-priced cigars. In 1921 there were 15,000 cigar factories, in 1947 only 822. The output

of the industry has remained almost unchanged through the decades, while cigarette production followed a steeply rising curve. The two dominating cigar manufacturing centers in the United States are Tampa, Florida, and Philadelphia. An American duty imposed on cigars from Havana made it feasible to make "Havana cigars" inside the United States in nearby Florida, employing chiefly Cuban and Spanish labor and using tobacco imported from Cuba.[2] With increased mechanization the North, chiefly Philadelphia, has gained in relative importance. The Northern states lead in the domestic growing of raw materials for the cigar industry. Pennsylvania has the largest production of filler tobacco, Wisconsin is a pre-eminent producer of binders, and the Connecticut Valley of wrappers.

## Rubber

Rubber manufacturing[3] has followed a production curve which is strongly correlated with that of the automotive industry. But rubber production has a pre-history which was of significance for the present distribution pattern of the industry. Following the discovery of rubber vulcanization by Goodyear in 1839, small factories began to spring up in New England making rubber boots and shoes. One of these early rubber manufacturers, Goodrich of Hastings, New York, was financially encouraged by a friend to move his factory to Akron in 1870. As the small plant, manufacturing fire hose and wringer rolls, was successful, other companies started similar plants in Akron. Among products turned out in this new rubber center at the turn of the century were pneumatic tires for the carriage industry of the Middle West.[4] When the automobile boom got under way about 1910, Akron was rather conveniently located to supply the car factories of Detroit with

1. E. B. Alderfer and H. E. Michl, *Economics of American Industry* (New York, 1950).

2. J. Russel Smith and M. O. Phillips, *North America* (New York, 1942).

3. *The rubber products industry* employed 259 thousand people in 1947 according to the Census of Manufactures.
*Tires and inner tubes*, 116 thousand (Ohio 50, Mich. 12, Calif. 11, Pa. 8).
*Rubber footwear*, 28 thousand (Mass. 11, Conn. 5).
*Rubber industries*, n.e.c., 113 thousand (Ohio 33, N.J. 16, Mass. 10, Ind. 8).

4. Renner, Durand, White, and Gibson, *World Economic Geography* (New York, 1953).

tires and inner tubes. Akron rapidly developed into one of the most specialized manufacturing cities in the country.

During the last decades there has been a certain decentralization in the rubber industry, resulting chiefly from the need for economy in distribution. The four big rubber companies, Goodyear, Firestone, United States Rubber and Goodrich have acquired plants in several American states, as well as abroad. Los Angeles is the largest of these secondary centers. As California is the most remote of the big domestic tire markets and, consequently, as the biggest savings in freight costs will be made if a branch plant is located in this state, there was a tendency when one of the big rubber companies built a branch plant in California for the others soon to follow in order not to be outcompeted on the West Coast. This procedure has been common also in other industries dominated by big companies, and it has been an important factor behind the growth of Los Angeles as a manufacturing center. Los Angeles and San Francisco, by far the largest cities on the Pacific Coast, seem to have been the alternative locations for plants of this type. For several reasons— the center of population in California has been moving towards Los Angeles, and the latter city is more industrialized than San Francisco—Los Angeles has been chosen in most cases.

Among the secondary tire centers is Detroit. Plants have been built here to supply new cars more economically with their original set of tires. In the large Southern market Gadsden, Alabama, was chosen by one of the "Big Four" companies as a location for the largest single tire factory outside of Akron.

Most of the rubber footwear industry is concentrated in the shoe manufacturing district of New England, with Massachusetts as the leading state.

## Leather and Shoes

The leather and shoe industry[1] is still dominated by New England and the Middle Atlantic states, where this type of manufacturing first started. New England alone has about one-third of the employees in the shoe industry. According to Hoover,[2] four periods can be distinguished in the history of shoe making which are important for an understanding of the present distribution pattern.

In the first period, until about 1760, each village and neighborhood community produced its own supply of shoes, using local materials. In certain districts along the Atlantic seaboard, chiefly in eastern Massachusetts and around Philadelphia and New York City, population was dense enough for an elementary division of labor. Local craftsmen, at first itinerant and later established in their own shops, made shoes to measure for the rest of the community.

The next stage consisted in the division of shoemaking into operations which could be performed by home workers.[3] All three districts mentioned above took part in this development. Shoes became an article of trade. Conditions were especially favorable for the development of manufacturing in eastern Massachusetts, where many people were living in the coast towns, engaged in trading, shipping, fishing, and shipbuilding,

1. Manufacturing of *leather and leather products* in 1947 employed 383 thousand people, according to the Census of Manufactures.
   *Leather tanning and finishing*, 53 thousand (Mass. 12, Pa. 8, N.Y. 5, Wis. 5).
   *Footwear, except rubber*, 229 thousand (Mass. 44, Mo. 34, N.Y. 30, Ill. 18, Pa. 16, N.H. 16, Me. 14, Ohio 14, Wis. 11, Tenn. 9).
   *Handbags and purses*, 20 thousand (N.Y. 13).
   *Leather dress gloves*, 9 thousand (N.Y. 7).
2. E. M. Hoover, Jr, *Location Theory and the Shoe and Leather Industries* (Cambridge, 1937).
3. The *putting-out system* with its division of labor paved the way for mechanization in the shoe industry as well as in other industries, e.g. the clothing industry.

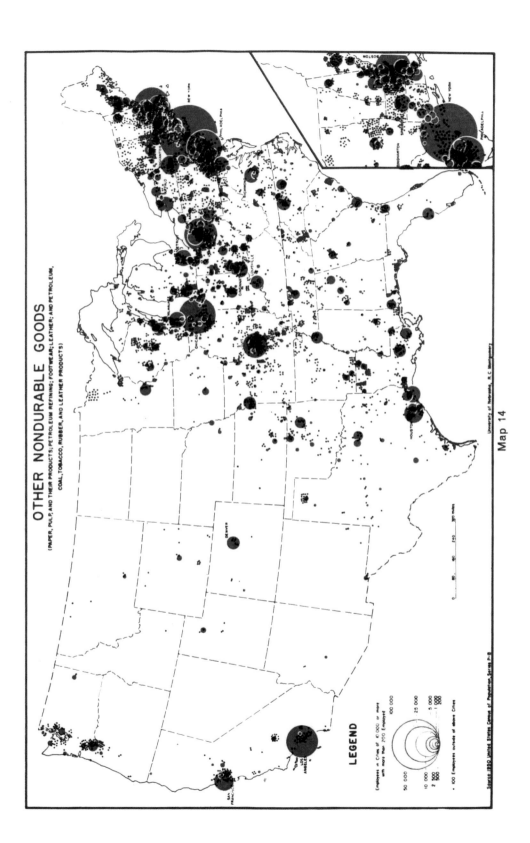

# OTHER NONDURABLE GOODS

(PAPER, PULP, AND THEIR PRODUCTS; PETROLEUM REFINING; FOOTWEAR; LEATHER; AND PETROLEUM,
COAL, TOBACCO, RUBBER, AND LEATHER PRODUCTS)

LEGEND

Employees in Cities of 10,000 or more
with more than 200 Employed

50 000

25 000

10 000

2 500

100 000

5 000

1 000

• 100 Employees outside of above Cities

0    80    160    240    320 miles

Source: 1950 United States Census of Population, Series P-3

University of Nebraska, R.C. Montgomery

Map 14

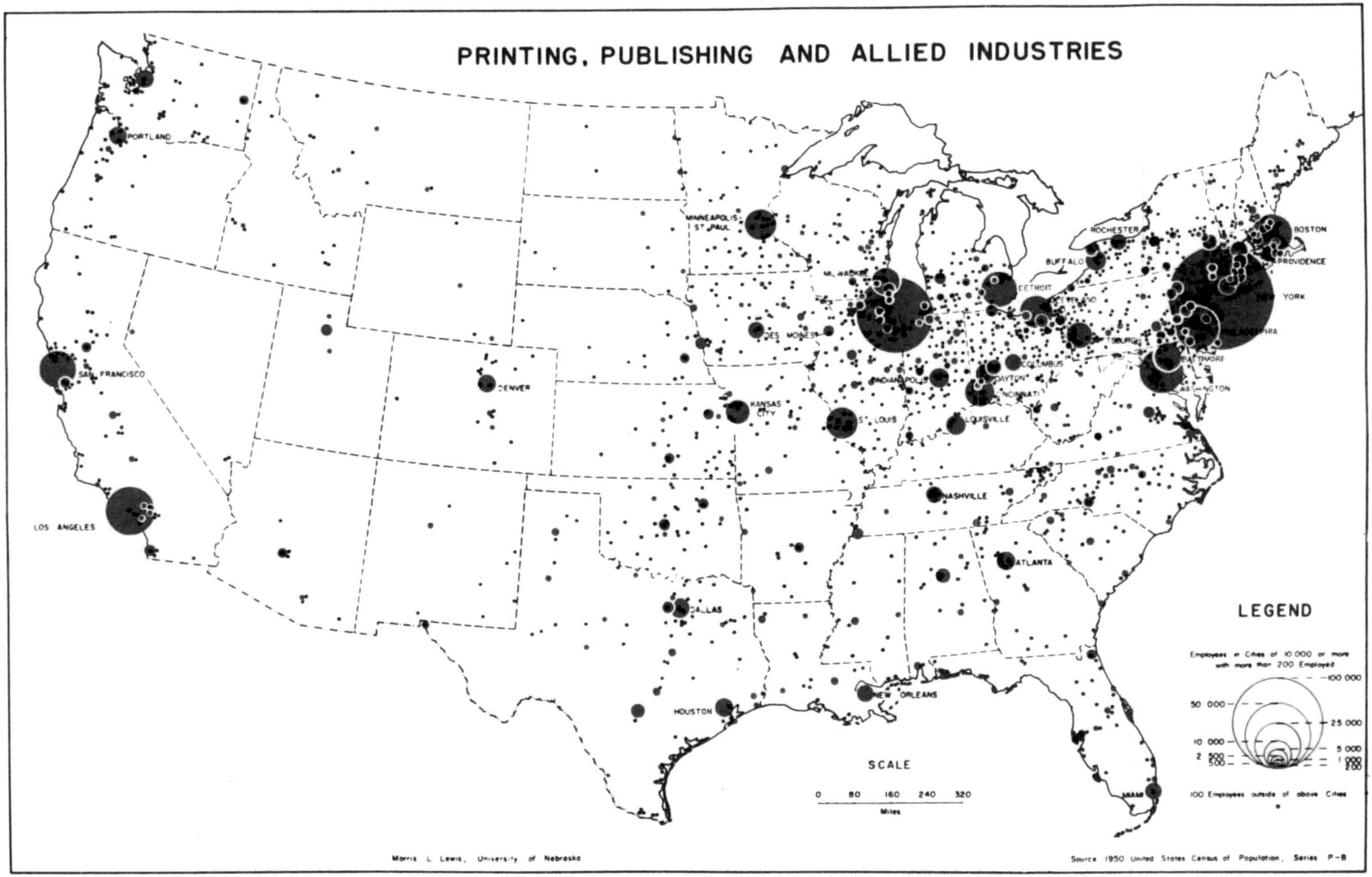
PRINTING, PUBLISHING AND ALLIED INDUSTRIES
PORTLAND
SAN FRANCISCO
LOS ANGELES
DENVER
MINNEAPOLIS
ST. PAUL
DES MOINES
KANSAS CITY
DALLAS
HOUSTON
MILWAUKEE
ST. LOUIS
INDIANAPOLIS
LOUISVILLE
NASHVILLE
NEW ORLEANS
DETROIT
CLEVELAND
COLUMBUS
DAYTON
CINCINNATI
ATLANTA
ROCHESTER
BUFFALO
PITTSBURGH
BALTIMORE
WASHINGTON
PHILADELPHIA
NEW YORK
BOSTON
PROVIDENCE
MIAMI
SCALE
0 80 160 240 320
Miles
LEGEND
Employees in Cities of 10 000 or more with more than 200 Employed
100 000
50 000
25 000
10 000
5 000
2 500
1 000
500
200
100 Employees outside of above Cities
Morris L. Lewis, University of Nebraska
Source 1950 United States Census of Population, Series P-B

Map 15

and where agriculture was of relatively less importance than elsewhere on the coast or in the interior because of the poor soils. Capital was accumulated in the towns, and the urbanization along the coast meant a concentrated market as well as a supply of labor. Fishermen's and sailors' wives and daughters were the labor supply for the first localized shoe industry.

The third period, from about 1860, is marked by the introduction of the sewing machine, which led to a concentration of production in factories and to reduced requirements of labor skill. It became possible to establish shoe factories nearer the fast growing regions in the Middle West. In the beginning such new shoe centers as Cincinnati, Chicago, and St Louis made chiefly medium- or low-grade shoes, but as labor gained in experience they approached eastern factories in quality.

From about the turn of the century a stronger emphasis on style became a factor of relevance for the location of the shoe industry. The increasing importance of fashion changes in shoes helped to preserve the specialized, moderate-sized plant as a characteristic of the industry[1] and it also strengthened the market orientation.

Among shoe centers in the old district of New England, Lynn is probably the best known. It is now a suburb of Boston. With many shoe factories in other suburbs also and in the city proper, Boston is a shoe city of C-type. Shoe manufacturing is for Boston what the apparel industry is for New York. Brockton and Haverhill in eastern Massachusetts, Manchester in New Hampshire and Auburn - Lewiston in Maine are other important shoe towns in the dominant New England district.

Binghamton, New York, has one of the world's largest companies in the shoe and tannery industry, the Endicott-Johnson Shoe Company. The success of this enterprise is due more to business genius than to any locational advantage.[2] The company has a large number of factories, specializing in different types of shoes, in the three cities of Binghamton, Endicott and Johnson City which together constitute the Binghamton urbanized area. Southeastern Pennsylvania, with Philadelphia as its center, the New York urbanized area, and Rochester, New York, are other important concentrations of shoe manufacturing plants in the eastern part of the Manufacturing Belt.

In the western part of this belt the big cities of St. Louis, Chicago, Milwaukee, and Cincinnati are important shoe centers. In all these cities shoe manufacturing on a large scale got under way in the decades following the Civil War, or in the years when the Irish and the Germans dominated the immigration to the United States. *Hoover* points out that these cities, as well as Rochester, New York, had a conspicuous German element. The principal cause of the German emigration at this time seems to have been an agricultural crisis in districts of small holdings in southwestern Germany. People coming from these areas were trained in certain trades, among them the tanning and working of leather. Wherever the Germans went they carried tanning and shoemaking with them. Many had savings and could afford to look around for the most promising places. The Middle West as a whole, and especially the above-mentioned cities, were rapidly expanding at this time.[3] The migration of the shoe industry from New England has been fairly analogous to

1. An influence in the same direction is the policy of leasing and not selling machinery which has been adopted by the monopolistic United Shoe Machinery Corporation. Royalty is paid per unit of output regardless of the location of the factory or the amount of use of the machine. It has required relatively small financial backing to start a shoe factory.

2. C. L. White and E. J. Foscue, *Regional Geography of Anglo-America*. (New York, 1943).

3. Hoover, *op. cit.*, pp. 223 ff.

what took place in the cotton textile industry: first only the coarser grades but eventually the full line of products.

Shoe making is by far the most important leather industry. Of the less important branches mention shall only be made of the concentration of glove manufacturing to Gloversville and Johnstown in the Mohawk Valley of New York. This specialization goes back to the eighteenth century, when a group of Scottish glovemakers settled here. For a long time glove making was a home industry before it was transferred to factories.[1]

## Ubiquitous Manufacturing Industries

CONSTRUCTION

At first it may seem inappropriate to use the term city forming industry in reference to construction. It may seem too much like a new variant of the old story about the inhabitants of a town washing each other's shirts for a living. Two points, however, should be borne in mind.

1. Building contractors often operate on a large scale, building homes, factories, roads, etc., even outside of their home towns. If they use the city as a base of operation for their machinery and skilled workers, "house and road manufacturing" is just as clearly a city forming activity as the making of automobiles.

2. Cities grow at different rates. If only the construction of new homes is considered, a city increasing its population by, say, four per cent annually should have, all other things being equal, roughly twice as many people employed in construction as a city growing at a rate of two per cent per year. Obviously the contractors build houses for the cities' own populations in both cases and their production should be considered as city serving according to the terminology used here (page 15). Cities with a high growth rate are, however, not "normal"; a large part of their expansion will undoubtedly be paid for by savings made in other parts of the country. Their supernormal construction percentages will be maintained only as long as they maintain their high growth rates. The extreme case is that of a city being built from scratch. During the construction period that agglomeration is an extremely one-sided construction town; construction is the *only* city forming activity.

In all cities construction is a major industry. Only in about ten per cent of the cities are less than four per cent of the gainfully employed population engaged in construc-

1. J. R. Smith and M. O. Phillips, *North America.* (New York, 1942), p. 162.

TABLE 3. Regional Differentiation of Construction as a City Forming Activity.
    I–X Construction, decils. See Fig. 17.

| Zones | I | II | III | IV | V | VI | VII | VIII | IX | X | No. of cities |
|---|---|---|---|---|---|---|---|---|---|---|---|
| Northeast . . . . . . . . | 23 | 20 | 13 | 18 | 18 | 7 | 4 | 4 | 1 | 0 | 108 |
| East Manufacturing . . . . | 30 | 27 | 23 | 13 | 5 | 6 | 1 | 1 | 2 | 0 | 108 |
| West Manufacturing . . . . | 18 | 22 | 23 | 16 | 13 | 10 | 5 | 1 | 0 | 0 | 108 |
| North–Central . . . . . . | 11 | 10 | 13 | 20 | 17 | 10 | 8 | 8 | 8 | 3 | 108 |
| Southeast . . . . . . . . | 1 | 4 | 10 | 10 | 19 | 18 | 22 | 10 | 7 | 7 | 108 |
| South . . . . . . . . . . | 0 | 2 | 1 | 5 | 3 | 16 | 13 | 13 | 24 | 31 | 108 |
| Prairie . . . . . . . . . | 0 | 0 | 1 | 2 | 7 | 13 | 20 | 24 | 20 | 21 | 108 |
| West . . . . . . . . . . | 3 | 1 | 3 | 2 | 5 | 6 | 14 | 25 | 25 | 24 | 108 |
| No. of cities . . . . . . . | 86 | 86 | 87 | 86 | 87 | 86 | 87 | 86 | 87 | 86 | 864 |

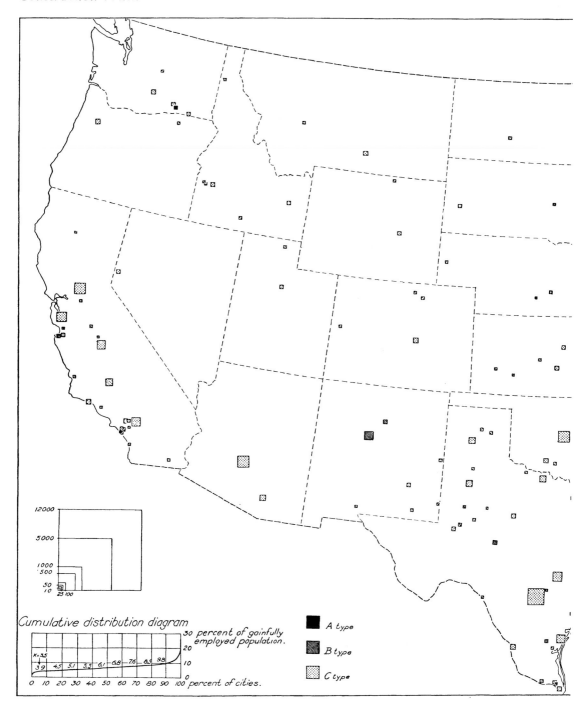

FIG. 16. Construction is an important industry in the fast-growing cities of the western, central and southern parts of the United States.

TABLE 4. Correlation Table: Construction / Manufacturing, Including Mining and Construction. I–X, Construction, decils. A–J, Manufacturing, decils. For calculation of decils, see page 14. and the diagrams of Map 1 and Figure 16.

|   | I | II | III | IV | V | VI | VII | VIII | IX | X | No. of cities |
|---|---|----|-----|----|---|----|-----|------|----|---|---------------|
| A | 4 | 0 | 3 | 7 | 5 | 11 | 15 | 15 | 18 | 8 | 86 |
| B | 1 | 1 | 3 | 8 | 5 | 6 | 12 | 14 | 16 | 21 | 87 |
| C | 0 | 1 | 2 | 6 | 7 | 6 | 8 | 23 | 17 | 16 | 86 |
| D | 2 | 2 | 4 | 8 | 9 | 15 | 15 | 9 | 9 | 14 | 87 |
| E | 2 | 3 | 6 | 3 | 9 | 16 | 19 | 13 | 10 | 5 | 86 |
| F | 2 | 6 | 5 | 15 | 15 | 8 | 7 | 9 | 8 | 12 | 87 |
| G | 5 | 14 | 17 | 10 | 14 | 9 | 7 | 0 | 5 | 5 | 86 |
| H | 8 | 22 | 21 | 13 | 10 | 7 | 1 | 2 | 1 | 2 | 87 |
| I | 19 | 21 | 15 | 10 | 8 | 7 | 2 | 0 | 3 | 1 | 86 |
| J | 43 | 16 | 11 | 6 | 5 | 1 | 1 | 1 | 0 | 2 | 86 |
| No. of cities | 86 | 86 | 87 | 86 | 87 | 86 | 87 | 86 | 87 | 86 | 864 |

tion; about the same number of cities have a percentage of ten or more (Fig. 16).

There is no correlation between city size and construction percentage. Both large and small cities are evenly distributed on the construction decils. Cities west of the Mississippi and in the South have a high construction percentage, those in the Manufacturing Belt have a low one (Table 3). Cities with a high manufacturing percentage

have a low construction rate, and vice versa (Table 4). Fast growing cities in general have a higher construction percentage than slow growing ones (Table 5).[1] The observations from the three tables can be summed up in the following way: the fast-growing cities west of the Mississippi and in the South, low in manufacturing, have a higher construction percentage than the slow-growing manufacturing cities of the Manufacturing Belt.

In 1950 the only construction towns of A-type in the United States were the twin cities of Pasco-Kennewick, Washington, near the atom city of Richland. Obviously, this unique industrial structure must be ephemeral.

There were fourteen construction cities of B-type: Monterey and Costa Mesa, California; Albuquerque, Santa Fe and Las Cruces, New Mexico; Lamesa, Midland, San Angelo, Alice and Victoria, Texas; Oak Ridge, Tennessee; Vicksburg, Mississippi; Fort Lauderdale and Lake Worth, Florida. In eleven of these cities construction was the leading city forming activity and the only one of B-type. In several instances the high construction percentage can, at least

1. Data on population growth between 1940 and 1950 are available for 850 urban agglomerations as defined on pages 22 ff. No growth data are given in the census publications for unincorporated places. For cities with 50,000 inhabitants and more in 1940, growth figures refer to Standard Metropolitan Areas. The 1940 census did not use the concept of urbanized area. For a correlation of city growth and construction percentage a comparison of the population increase between, say, 1948 and 1951 would have been more relevant, but such data were not available. In 1950 college students were for the first time included in the population of the communities where they attended school, resulting in an apparent population increase for all university cities. There are several other pitfalls in comparing city populations at two different times. Some of the population increase may have taken place in suburbs outside of the administrative city limit (in cities of 10 to 50 thousand inhabitants); earlier such population growths may have been realized by incorporations, giving a false picture of when the increase actually took place, etc. Table 5 should thus not be pressed on details, but the large number of observations should ensure a safe basis for general conclusions.

TABLE 5. Correlation Table: Construction / Population Growth, 1940–50.[1]
I–X, Construction, decils. A–J Population Growth, decils. For calculation of decils, see page 14, Figure 16 and the diagram of Map 17.

|   | I | II | III | IV | V | VI | VII | VIII | IX | X | No. of cities |
|---|---|----|-----|----|---|----|-----|------|----|---|---------------|
| A | 27 | 16 | 9 | 8 | 9 | 7 | 3 | 4 | 1 | 1 | 85 |
| B | 13 | 14 | 18 | 14 | 10 | 5 | 3 | 3 | 4 | 1 | 85 |
| C | 10 | 13 | 14 | 15 | 10 | 3 | 8 | 8 | 2 | 2 | 85 |
| D | 10 | 19 | 13 | 10 | 8 | 12 | 5 | 3 | 4 | 1 | 85 |
| E | 8 | 8 | 10 | 11 | 14 | 14 | 11 | 4 | 3 | 2 | 85 |
| F | 3 | 9 | 8 | 13 | 8 | 9 | 11 | 8 | 11 | 5 | 85 |
| G | 2 | 2 | 8 | 3 | 7 | 12 | 12 | 17 | 14 | 8 | 85 |
| H | 1 | 3 | 2 | 4 | 7 | 8 | 11 | 21 | 16 | 12 | 85 |
| I | 1 | 2 | 3 | 7 | 7 | 7 | 13 | 13 | 13 | 19 | 85 |
| J | 3 | 1 | 1 | 0 | 6 | 8 | 10 | 5 | 18 | 33 | 85 |
| No. of cities | 85 | 85 | 85 | 85 | 85 | 85 | 85 | 85 | 85 | 85 | 850 |

partly, be referred to conditions specific for the city: Oak Ridge is a newly built atom city; Vicksburg is the headquarters of the U.S. Mississippi River Commission in charge of large flood-control works in six states along the lower Mississippi; the Texan cities are located in large oil-fields; Albuquerque has large military installations and Santa Fe is near the atom town, Los Alamos, where the first atomic bomb was made. In most of these cities, however, as well as in most construction cities of C-type, the high construction percentage seems to be chiefly the result of a high growth rate.

PRINTING AND PUBLISHING

In the printing and publishing industry there are both sporadic and ubiquitous branches.[2] Books are almost exclusively published and to a large extent also printed in a few of the largest cities. The same holds true of periodicals. Newspapers, on the other hand, are published in practically every American town with more than 10,000 inhabitants and are one of the most ubiquitous manufacturing branches in the United States.[3] Almost every city also has

one or more establishments for commercial printing. The last two branches are by far

1. The correlation between population growth and employment in construction would have been even greater but for the many college towns with their seemingly rapid growth and often low construction percentage. Most of the cities in the lower left corner of the table are of this type.
A conspicuously large number of cities in the Prairie zone and some in the West zone have a low growth rate but a relatively high construction percentage (compare Table 3 and Map 17). Many of the cities in the upper right hand corner of Table 5 are of this type. Can these high construction percentages, combined with low growth rates, be explained by the exceptional importance in their zones of irrigation works, transcontinental highway construction, etc.?
2. *Newspapers* in 1947 employed 234 thousand people, according to the Census of Manufactures (N.Y. 32, Calif. 19, Pa. 18, Ill. 17, Ohio 13, Mass. 11, Tex. 10, Mich. 9).
*Commercial printing*, 193 thousand (Ill. 37, N.Y. 33, Pa. 14, Ohio 14, Calif. 11, Mass. 8, N.J. 8).
*Periodicals*, 69 thousand (N.Y. 26, Ohio 9, Pa. 8, Ill. 7).
*Books, publishing and printing*, 40 thousand (N.Y. 22, Ill. 6).
*Books, printing*, 11 thousand (N.Y. 4).
*Litographing*, 52 thousand (N.Y. 12, Ill. 7).
*Bookbinding*, 22 thousand (N.Y. 9).
3. Newspaper plants, bakeries, artificial-ice plants and soft-drink bottling plants operate in more than a third of the nearly 3,000 counties of the United States. For a complete tabulation of all 1935 census industries by number of establishments and number of counties see U.S. National Resources Planning Board, "Industrial Location and National Resources." Government Printing Office (Washington, 1943).

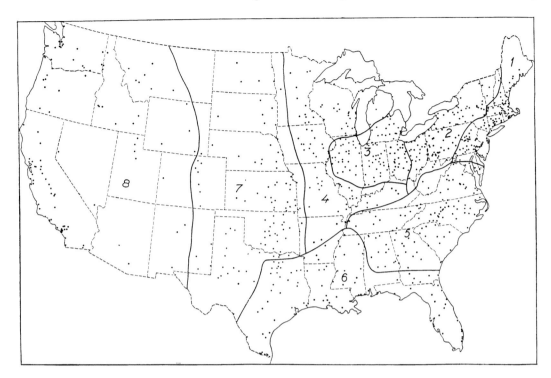

FIG. 17. Map showing the limits of the eight zones which have been used for the analysis of the regional differentiation of percentages for some industries. All zones have 108 cities each. See tables 3, 8 and 12. The eight zones are: 1 Northeast, 2 East Manufacturing, 3 West Manufacturing, 4 North-Central, 5 Southeast, 6 South, 7 Prairie, and 8 West.

the leading ones, accounting for about 60 per cent of the total employment.

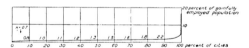

FIG. 18. Cumulative distribution diagram for printing and publishing.

The adjustment of the newspaper industry to the general consumer market seems to be more intimate in the United States than in other countries, e.g. Sweden, a fact which may reflect a difference of degree in the functions of newspapers in different countries. In the United States about two-thirds

1. J. G. Glover & W. B. Cornell, *The Development of American Industries* (New York, 1946).
2. Olof Hedbom, *Sveriges grafiska industri*. Industriens upplysningstjänst (Stockholm, 1949).

of all newspaper revenues come from advertising.[1] The corresponding figure is about one-half for Swedish newspapers.[2] Advertisments cover a larger portion of available space in American papers than in European ones. The newspapers of the big cities, New York, Chicago, Los Angeles, Washington, etc., do not have so large a portion of their circulation outside of their respective metropolitan areas as the papers of Stockholm and other big cities in Europe. The Stockholm papers, air-borne to strategic points and from them distributed by surface transport, have a nationwide circulation. A relatively large part of Swedish households even in remote areas subscribe to a Stockholm paper in addition to their local paper. A similar situation is also true of the United Kingdom and the London papers. With the greater importance in the United States of newspapers as carriers of advertisments in general and especially those of firms with a geographically restricted market, the small-city papers hold a rather strong position. Although the American metropolises have a larger newspaper employment than would correspond to their population, big-city papers are not so dominant in the United States as they are in Europe.

New York is the leading publishing center for books and periodicals. The city gradually arrived at this position during the nineteenth century, when the European influence on American book publishing was stronger than at present. In 1820 about 70 per cent of all the books published in the United States were of foreign origin, but the proportion gradually declined until, by the middle of the century, 80 per cent were by American authors.[1] New York emerged as the publishing center in competition with Boston and Philadelphia, which still are the homes of important publishing firms. In the interior, Chicago became the leading center, second only to New York. Many of the books published in the big cities mentioned are printed in small towns with lower production costs.

Washington, D.C., of little importance in most manufacturing industries, stands out as a leading printing and publishing center. It is the chief news producing spot in the United States, where all newspapers of any consequence must be represented. The Government Printing Office in Washington is perhaps the largest single printing and publishing establishment in the world.

There are two American printing towns of B-type, Clinton, Massachusetts, and Crawfordsville, Indiana, and seven of C-type, Concord, New Hampshire, Franklin, Pennsylvania, Ashland, Athens and Springfield, Ohio, Menasha-Neenah, Wisconsin, and Kingsport, Tennessee.

## FOOD PROCESSING

### Meat Packing

The meat packing industry employs more people than any other branch of food processing.[2] The money which the meat packers pay to farmers for livestock makes up about 25 per cent of the total farm income.[3]

In the early phase of the industry livestock was raised on the Atlantic seaboard, and the processing for consumption was accomplished by local butchers. As Eastern cities grew, supplies had to be brought from points farther and farther away. Livestock from distant localities had to be driven across the country, but with the advent of railroads the animals were shipped in stock cars. During the middle of last century the traffic in live animals from settlements in the Ohio and Mississippi valleys to the Atlantic seaboard became important. Shipment of meat products over long distances was confined entirely to cured meat, chiefly pork. It was salted and packed in barrels, and thus the name "packing industry" came into existence. Cincinnati, with the nickname "Porkopolis," was the largest center by 1850.[4]

The introduction in 1870 of the refrigerator car made it possible to shift the industry to the source of livestock. This was the logical location since only about half the weight of the animal, in the case of cattle, could be sold as meat. The result was a rapid development of the meat packing industry in Chicago and other Middle Western cities. The boom in Chicago had already started during the Civil War with the pack-

1. Glover and Cornell, *op. cit.*
2. *Meat packing* in 1947 employed 208 thousand people, according to the Census of Manufactures (Ill. 29, Iowa 23, Minn. 14, Kans. 13, Tex. 10, Calif. 10, Nebr. 9, Mo. 9, Pa. 9, Ind. 9).
3. E. B. Alderfer and H. E. Michl, *Economics of American Industry* (New York, 1950).
4. J. G. Glover and W. B. Cornell, *The Development of American Industries* (New York, 1946).

ing of pork, but the refrigerator car firmly established the city as the leading center of meat packing in general. Swift, a New England butcher, and Armour, a Milwaukee pork packer, established their companies in Chicago. They are today the two leading concerns, with hundreds of plants in the United States and abroad. The "Big Four" interstate packers, including also Wilson and Cudahy, control almost half of the total packing business.

The plants of Chicago, the rail focus of the Middle West with one of the world's largest concentrations of major railroad lines, bought cattle from the expanding Western ranges and cattle and pork from the rising Corn Belt centered in Iowa. The finished product was shipped in all directions but chiefly to the big Eastern market. Chicago still is the greatest single meat packing center, but it no longer dominates the industry as it once did. An increasing percentage of livestock is shipped by truck which lessens the transportation advantage of Chicago. In recent decades meat packing has been migrating farther west, and centers closer to the producing areas have gained in relative importance. Such centers are St. Paul, Sioux City, Cedar Rapids, Omaha, St. Joseph, Kansas City, St. Louis, Oklahoma City, Fort Worth and Denver. About two-thirds of American livestock is raised west of the Mississippi River, whereas more than two-thirds of the population lives east of this river. Thus, either livestock or meat, or both, must move considerable distances. The tendency is to minimize the movement of the livestock. The packing industry in contrast to most other industries had a

considerable increase in the number of establishments during the 1940's, partly reflecting a decentralization of the industry from the Middle West.

The rapid growth of population in California has made Los Angeles and San Francisco important meat packing centers. Although livestock is the leading source of income for Californian farmers, worth more than vegetables and fruits together, the state is a deficit region. The supply area of California has been pushed further and further east into areas which used to ship their animals to Midwestern centers.

In the East, especially in Pennsylvania and New York, local packers with small establishments are a characteristic of the industry. This is due chiefly to the large kosher demand in New York and other large Eastern cities.

## Flour Milling

Flour milling, one of the oldest manufacturing industries in the United States, is a rather unimportant industry from the point of view of employment.[1] It is highly mechanized and has a low value added. Labor costs only make up three per cent of the sales value of the American flour mills, as compared to 70 per cent for direct raw materials.

The milling industry migrated with the population to the interior parts of the continent. In the later part of the colonial period Philadelphia was the leading center. About the middle of last century Rochester, New York, became a prominent milling town, partly because of the Erie Canal, and at the same time St. Louis emerged as the chief center of the West because of her good transport facilities to the cotton-belt market in the South. The introduction of a new milling process from Hungary at the end of the nineteenth century, which made possible the processing of hard wheats,

---

1. *Flour and meal milling* in 1947 employed 39 thousand people, according to the Census of Manufactures (Kans. 4, Minn. 4, N.Y. 3, Ill. 3, Mo. 3, Tex. 3).

*Prepared animal feeds*, 55 thousand (N.Y. 5, Cal. 5, Ill. 5, Ohio 3, Mo. 3, Tex. 3).

*Cereal preparations*, 11 thousand (Mich. 5).

stimulated the growth of Minneapolis as the mill center of the extensive spring wheat area of the northern prairie region. Minneapolis for many years was regarded as "the flour city," producing the best breadbaking flour in the country. The yields of the spring wheat region declined, however, in the inter-war period owing to the repeated sowing of wheat year after year. Because of this and other factors, e.g. changes in transportation rates which adversely affected Minneapolis but were beneficial to Buffalo, the former city declined in importance.[1]

Buffalo, since 1930 the leading mill center in the world, has the advantage of proximity to large Eastern markets and cheap water transportation for the wheat.[2] It is located in a dairy region, which affords an excellent market for the by-product mill feed. Kansas City has grown as the milling center of the expanding winter wheat region and now equals Minneapolis in production.

## Bakeries

The baking industry, one of the largest food processing industries, employing about 200,000 people, is ubiquitous in its distribution. The biscuit and cracker branch, with almost one-fifth of the total sales in the industry, is dominated by three big companies. It is characterized by large plants located in the principal centers of population.

## Sugar Refining and Confectionary

Domestic production of cane and beet sugar[3] accounts for only a small share of the total American sugar consumption. In recent decades this share has always been less than 30 per cent. The largest source of American sugar is Cuba, but large quantities come also from the Philippines and from the American insular possessions, Hawaii, Puerto Rico, and the Virgin Islands.

Domestic production comes from two sources: cane sugar from the Gulf Coast and beet sugar from California and the Rocky Mountain states.

With these premises it is to be expected that cane sugar refining should be located in the big port cities. New York, Boston and Philadelphia in the Northeast, New Orleans in the South and San Francisco in the West are the leading centers. Beet sugar is produced in the beet growing regions, chiefly in California and Colorado.

Beet sugar and cane sugar do not normally compete on a national scale. The sugar beet can be grown over wide areas in the United States, but production is chiefly concentrated inland, except in the case of California. Most of the beet sugar is consumed in the interior parts of the country with little competition from cane sugar, which has to carry heavy transportation costs from the coastal refineries.[4] Government intervention in some form has affected the sugar industry since 1789, when the first tariff duty was imposed. Close government control still is a characteristic feature of the sugar refining industry in the United States, as well as in many other countries.

Manufacturing of confectionary products

1. E. B. Alderfer and H. E. Michl, *Economics of American Industry* (New York, 1950).
2. To this should be added the milling-in-bond privilege which can be economically made use of only by cities in the line of transportation followed by Canadian grain. It is offered by the U.S. Government and means that the wheat is milled in transit to foreign markets. The mill has to pay the salaries of government inspectors, and the flour cannot be sold in the United States. (White and Foscue, *Regional Geography of Anglo-America*, 2nd ed. [New York, 1954], p. 46.)
3. *Raw cane sugar*, 5 thousand people employed in 1947, according to the Census of Manufactures.
*Cane-sugar refining*, 17 thousand (N.Y. 4, La. 4).
*Beet sugar*, 13 thousand (Calif. 3, Colo. 3).
*Confectionary products*, 75 thousand (Ill. 16, N.Y. 10, Pa. 10, Mass. 7, Calif. 5).
*Chocolate and cocoa products*, 10 thousand (Pa. 5, N.Y. 3).
*Chewing gum*, 7 thousand (N.Y. 3).
4. E. B. Alderfer and H. E. Michl, *Economics of American Industry* (New York, 1950).

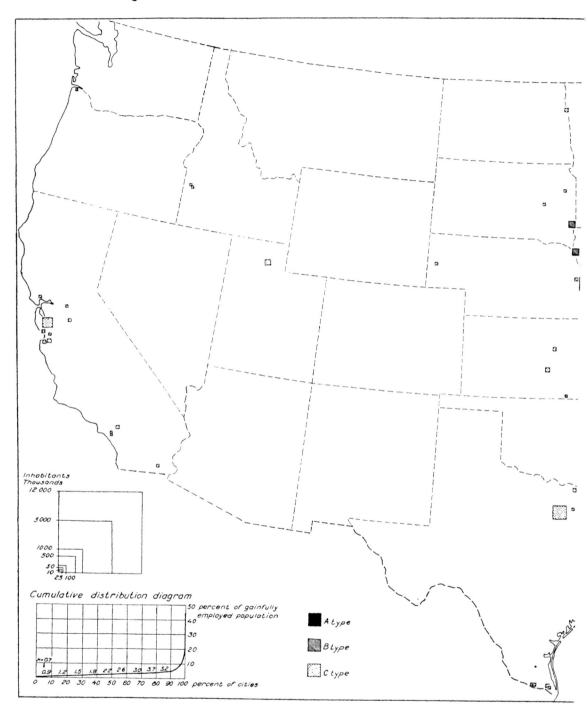

FIG. 19. The largest concentration of food manufacturing towns in the United States is found in the Corn Belt. Omaha, Nebraska, is the largest town of B-type.

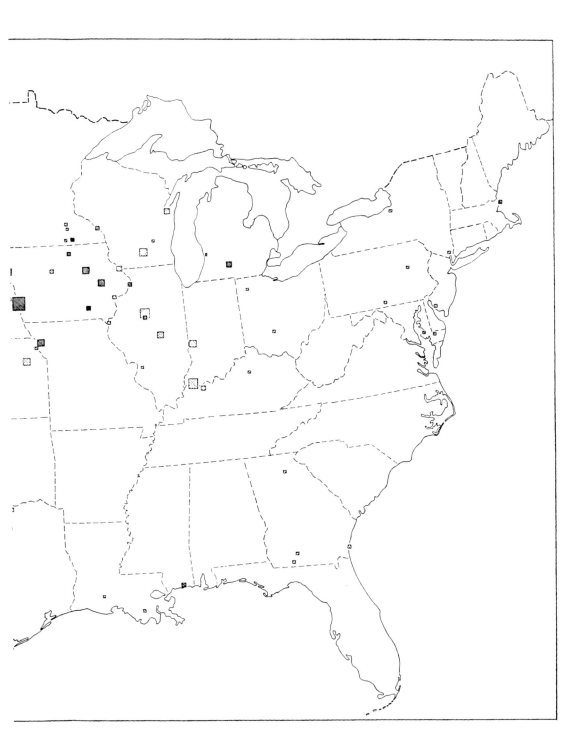

is scattered over the continent with a heavy concentration in the densely populated Manufacturing Belt.

## Canning

The canning industry[1] has grown more rapidly than the other major food industries during the first half of this century. It is highly seasonal, with employment figures varying between somewhat over 50,000 in the winter and a peak of 350,000 in September.[2] In most areas, housewives make up from one-half to two-thirds of the labor force. Canning is a typical example of a raw-material-oriented industry. The close relation between farm and cannery is reflected in the fact that a large share of the requirements of the canneries is grown and sold on a contract basis. In many areas receipts from the canneries make up a considerable share of the total farm income.

The modern canning industry builds on principles presented by the Frenchman Nicolas Appert in 1810. The method was immediately put into commercial practice in France and England and soon also in the United States. The first small canneries sprang up on the Atlantic seaboard about 1820, packing marine products chiefly but supplementing them by fruits and vegetables in season.[2] The years following the Civil War saw the first canneries in the Middle West and California. The rapid growth of commercial canning in recent decades was naturally stimulated by the urbanization of the Amerian population and by changes in

the habits of living bringing about increased reliance on prepared foods. The frozen food industry based on quick freezing is technically different from canning, but it performs the same function. It began about 1930, but the main development has been in the postwar period.

Canning plants are widely scattered over the continent, with California as the leading state. Canneries usually are rather small-sized enterprises. The typical plant employs an average of 50 to 60 people. As the seasonal variations in employment are great, however, the yearly averages understate the scale of operation. The reason for the small scale is the same as for the raw material orientation of the plants: the perishable nature of most of the raw materials. There are in the canning industry examples of large multi-plant financial organizations with plants in various parts of the country, but their relative importance is smaller than in most other industries. Biggest of these companies are California Packing Corporation, Heinz, and Libby, McNeill & Libby.

As can be seen from maps on the distribution of various fruit and vegetable crops in the United States, prepared by the Bureau of the Census,[3] most of the crops have a sporadic occurrence. Each crop tends to localize where climatic and soil conditions are most favorable. Even economic factors favor a localization of the crops. California leads in peach, plum, apricot, asparagus, and spinach canning. Sardines and tuna are also packed in California, shrimp in the Gulf states, beets and carrots in New York, pumpkins and squash in Indiana, and cherries in Michigan. The common vegetables—beans, peas, sweet corn, and tomatoes are grown and canned in many areas. The Atlantic Coastal Plain, especially the states of Maryland and New Jersey, is one of the most important vegetable canning areas in the United States (tomatoes, corn

1. *Canning and preserving, except fish*, in 1947 employed 136 thousand people, according to the Census of Manufactures (Calif. 26, Wis. 9, N.Y. 9, N.J. 8, Pa. 8, Md. 8, Ill. 8, Ind. 7, Fla. 6).
 *Canned sea food*, 20 thousand (Calif. 8, Me. 3).
 *Frozen foods*, 17 thousand.
2. E. B. Alderfer and H. E. Michl, *Economics of American Industry* (New York, 1950).
3. Available in several sources: see e.g. Van Royen, *The Agricultural Atlas of the World* (New York, 1954), or White and Foscue, *Regional Geography of Anglo-America* (New York, 1954).

and peas), and Baltimore is a leading canning center. Camden in the Philadelphia urbanized area has the largest plant for canned soups in the world. Another outstanding canning region dominated by common vegetables is the Agricultural Interior: Indiana, Illinois and Ohio (tomatoes), Minnesota (sweet corn) and Wisconsin (peas, green beans and cabbage for sauerkraut).[1]

The citrus districts of central Florida and the Lower Rio Grande Valley in recent years have developed a juice canning industry. Remarkable is the sudden appearance in the postwar years of frozen orange juice as a popular product on the American market. This type of quick freezing is largely a Florida business. More than half of the Florida orange crop, the largest in the country, is turned into frozen juice.

## Beverages

Manufacturing of beverages employs about 200,000 people in the United States.[2] The breweries and the soft drink plants make up the two dominant branches. Both have a more or less ubiquitous distribution. In the brewing industry there is a tendency towards larger and fewer establishments, but these are widely scattered over the country. In 1950 the five largest companies—each with one or a few plants—controlled almost 25 per cent of the beer business in the United States. Transportation costs for the bulky product and, in several cases, state legislation favor the small brewery serving its local market. The list of the leading brewery companies shows many German names. The fact that such "German" cities as Milwaukee and St. Louis are leading brewery centers can hardly be a pure coincidence.

The cheap soft drinks are even more sensitive to transportation costs than beer. Thus, to take just one example, although Atlanta is the home town of Coca-Cola and the essential syrup still is concocted in that city from a secret formula reputedly known to only three men,[3] the bottling plants are spread all over the world to minimize distribution costs.

Wine production is rather unimportant in the United States. It is highly localized in regions adjacent to the vineyards. California produces more than 80 per cent of the commercial output.[4] The distilling industry is also more sporadic than the brewing and soft drink industries. Whisky, accounting for 80 per cent of the total output, is an expensive product which can stand long transportation costs. Kentucky is the largest liquor producing state, with Pennsylvania ranking second. The many large distilleries in Kentucky, most of them near Louisville, use local and Midwestern grain, chiefly corn, to produce bourbon whisky. Water is an important locational factor, but tradition also plays a prominent role.

## Dairy Products

The preparation of milk for consumers and the making of butter, cheese and other milk products such as ice cream and condensed milk are today factory industries. In all, they employ about 200,000 people in the United States. Generally, they have a ubiquitous distribution, but the manufacturing of some dairy products is highly concentrated in the western part of the American Dairy Belt. The regional differentiation

1. White and Foscue, *op. cit.*
2. *Bottled soft drink plants* in 1947 employed 79 thousand people, according to the Census of Manufactures.
*Malt liquors*, 83 thousand (N.Y. 13, Wis. 9, Pa. 7, N.J. 6, Ohio 6, Mo. 6).
*Wines and brandy*, 8 thousand (Calif. 5).
*Distilled liquors, except brandy*, 30 thousand (Ky. 10, Pa. 5).
3. *Southeastern Excursion, Guidebook.* XVIIth International Geographical Congress (Washington 1952), p. 49.
4. E. B. Alderfer and H. E. Michl, *Economics of American Industry* (New York, 1950).

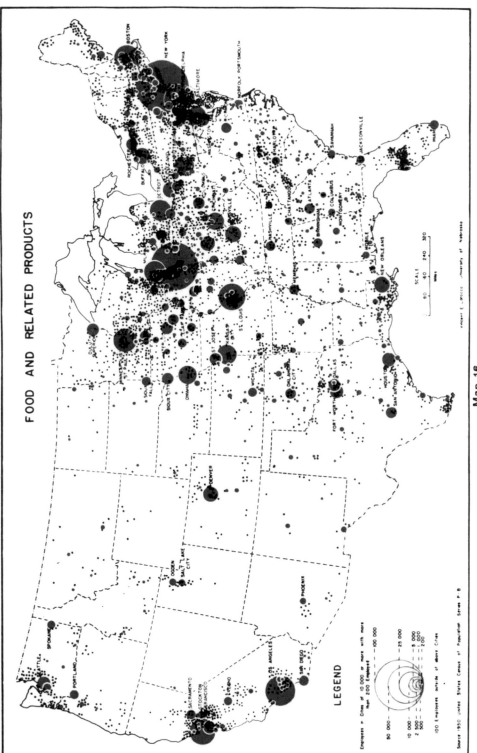

FOOD AND RELATED PRODUCTS

LEGEND

Employees in Cities of 10 000 or more with more
than 200 Employed

100 Employees outside of above Cities

SCALE

80    60    120    240    320

Miles

Source 1950 United States Census of Population Series P-8

Map 16

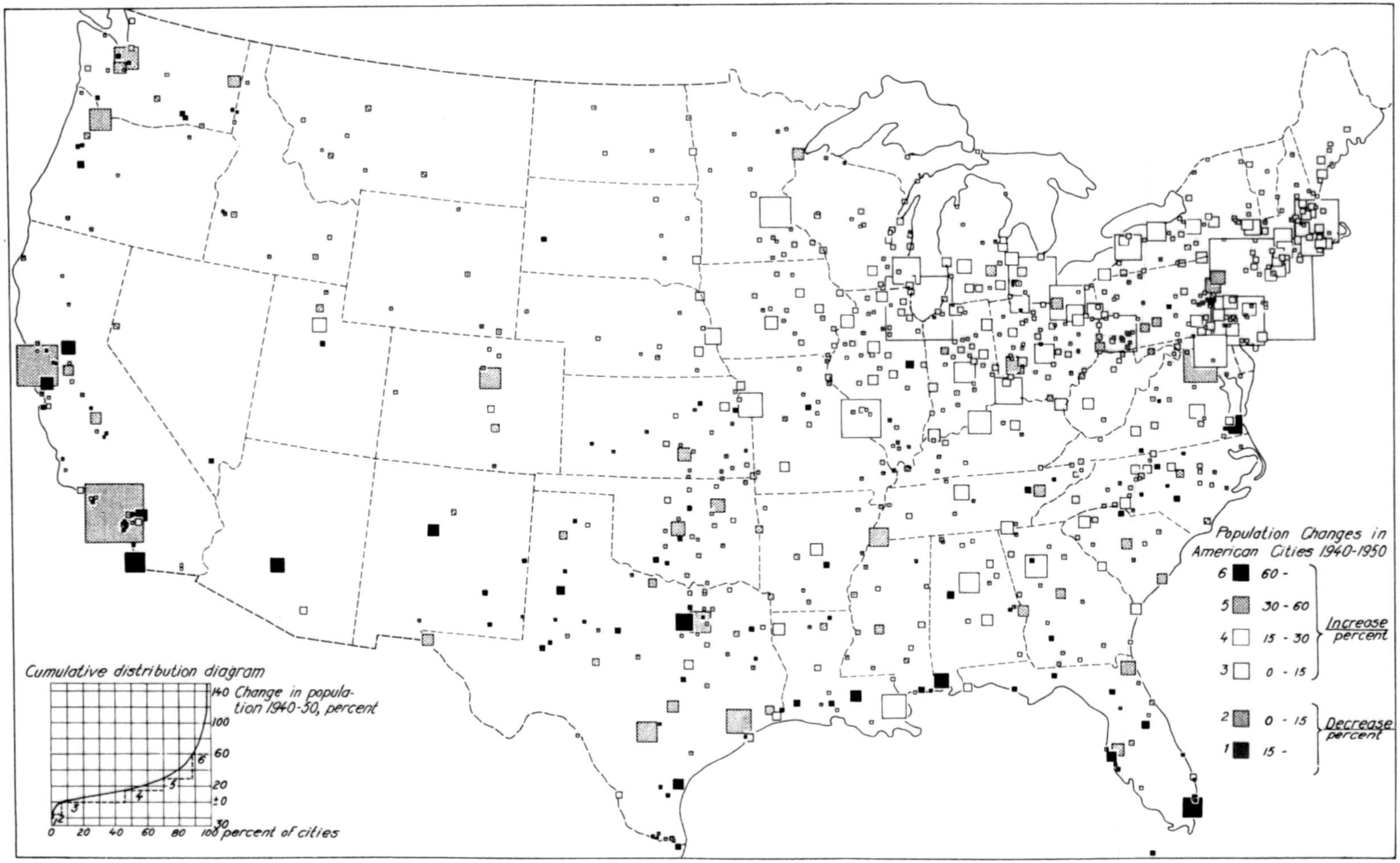
Population Changes in American Cities 1940-1950
6 60 -
5 30 - 60
4 15 - 30
3 0 - 15
Increase percent
2 0 - 15
1 15 -
Decrease percent
Cumulative distribution diagram
Change in population 1940-50, percent
140
100
60
20
±0
30
0 20 40 60 80 100 percent of cities

Map 17

in the dairy industry will best be understood against a knowledge of how the milk is utilized.

Milk is not shipped any farther than is absolutely necessary to obtain an adequate supply. It is low in value per unit of weight and it is perishable. Thus the immediate vicinity of any urban center in the United States, no matter where located, is a milk producing area. From this it follows that the highly urbanized Manufacturing Belt is an important dairy region. As the expanding city markets pay a higher price for milk than do creameries and cheese factories, such formerly specialized regions as northwestern and central New York (cheese) and northeastern Illinois (butter) have changed to milk production for the urban market. Today the western states of the Dairy Belt, Wisconsin and Minnesota, lead in the manufacturing of dairy products. The natural conditions for dairy farming here are very favorable, and the two states top the list of milk producing states.[1] The urbanization is lower than further east, resulting in a greater milk "surplus" than anywhere else in the United States. The

manufactured product, butter, cheese, etc., has a higher value per unit of weight and can therefore bear higher freight costs than milk. A few relatively inaccessible areas in the eastern portion of the dairy region, such as the part of New York north of the Adirondack Mountains, also specialize in manufactured dairy products. A considerable portion of the American production of butter and other dairy products comes from milk plants in the cities, which process their surplus milk.

The chief butter-producing area of the American dairy region is in eastern and southeastern Minnesota, northeastern Iowa and western Wisconsin. Creameries in the villages and small towns manufacture butter which is sold nationally. The area is the most concentrated source of butter in the nation.

Wisconsin is by far the leading cheese state, with about half of the total American production. Three areas in Wisconsin stand out as the most important: the Lake Michigan coast land north of Milwaukee, a band across north-central Wisconsin and the southwestern part of the state. Durand considers such human factors as national origins of population and an early start to be more important for an understanding of this regional specialization than the physical fact that all three areas have low summer temperatures.[2]

1. Wisconsin leads by a wide margin. In 1949 it produced 15.6 billion pounds of milk, followed by New York (8.7 billion pounds) and Minnesota (8.3 billion pounds). Wilcox, Krause and Brereton, "Utilization of Wisconsin Milk," Special Bulletin No. 3, Wisconsin State Department of Agriculture 1950.

2. Renner, Durand, White, and Gibson, *World Economic Geography* (New York, 1951).

# Service Industries

## Transportation and Communication Industries

The relative importance of different means of transportation in American intercity traffic may be gathered from the following figures:[1]

|  | Per cent of total ton-miles | Per cent of total passenger-miles |
|---|---|---|
| Railways . . . . . . . | 64 | 10 |
| Inland waterways, including Great Lakes | 19 | 0.6 |
| Pipe lines . . . . . | 10 | — |
| Highways . . . . . . | 7 | 89 |
| Airways . . . . . . . | 0.01 | 0.4 |

The figures refer to 1941. Since then pipe lines, highways and airways have in all probability increased their shares of freight transportation, but railroads still dominate the field. The relatively small railroad share of passenger traffic has probably decreased further in favor of the airways.

### RAILROADS

Railways in contrast to highway traffic, constitute one of the most important cityforming service industries. Railroad repair shops, division points where the trains change crews, classification yards, etc., have a sporadic occurrence. If they are located in small or medium-sized cities, the railroads will often be leading employers, and the cities railroad towns. Some of these towns have grown with the railroad. Pocatello, Idaho, for example, began as a collection of tents with the arrival of the Union Pacific in 1882.[2]

Chicago is by far the largest railroad center in the United States. It handles more railway traffic than that of the next two most important American railway centers, St. Louis and New York, combined.[3] Chicago's prominent position in the American railroad pattern stands out clearly on Ullman's map of the railroads of the United States, which are classified in six categories according to importance.[4] Railways are a

1. Harold M. Mayer, "Emerging Developments in Intercity Transportation," *The American Academy of Political and Social Science* (November, 1945), p. 53.
Mayer does not consider coastwise shipping. According to estimates made by the U.S. Interstate Commerce Commission, railroads in 1948 handled 54.5 per cent of total ton-miles of freight; coastwise shipping, Great Lakes shipping and inland waterways 28.9 per cent; pipe lines 9.5 per cent and highways 7.0 per cent. The figures are here cited after Edward L. Ullman, "Die wirtschaftliche Verflechtung verschiedener Regionen der USA," *Die Erde* (1955), p. 134.
2. White and Foscue, *Regional Geography of Anglo-America* (New York, 1943), p. 708. See also Robert Wrigley, Jr., "Pocatello, Idaho as a Railroad Center," *Economic Geography* (1943).
3. Harold M. Mayer and Allen K. Philbrick, "Chicago," *Industrial Cities Excursion, Guidebook.* XVIIth International Geographical Congress (Washington, 1952), p. 41.
4. Edward L. Ullman, "The Railroad Pattern of the United States," *Geographical Review* (1949), p. 244.

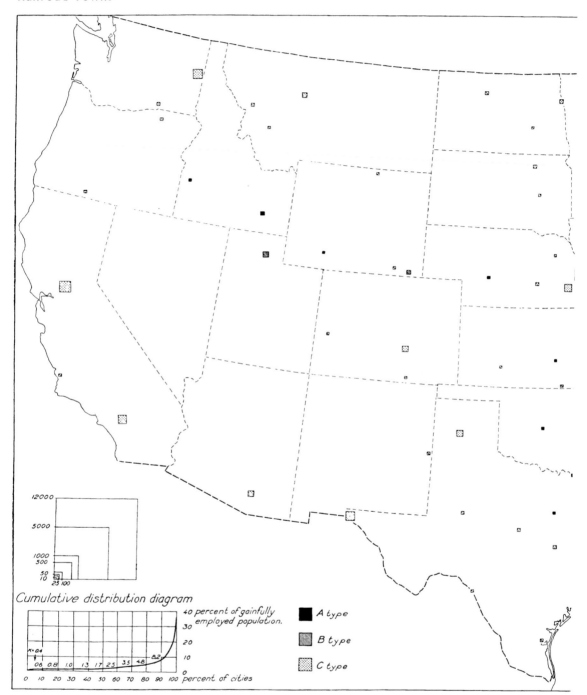

Cumulative distribution diagram

A type

B type

C type

FIG. 20. Railroad towns are common in the Manufacturing Belt with its dense net of railroads but they are also common in the wide belt of central United States, dominated by wholesale and retail towns, which is both a heavy-cargo generating area and a region traversed by trans-continental railroads. The largest railroad town of A-type in the United States is Altoona on the Pennsylvania Railroad.

major city forming activity in Chicago, but only one of many. Of the metropolises with more than one million inhabitants St. Louis and Pittsburgh have a higher percentage for railroad employment than Chicago (App. 2). St. Louis is the hub of many large railroads in the Middle West. It shows striking similarities to Chicago in its transportation functions. On Pittsburgh, the steel city, converge several important freight lines from Lake Erie ports with heavy traffic of iron ore and coal. But even more important is the Pennsylvania Railroad from Pittsburgh to New York with the heaviest traffic in America.[1]

Most of the railroad cities are located in the Manufacturing Belt with its dense railroad net and heavy traffic, and in the west-central parts of the Midwest, the states of Missouri, Kansas and Nebraska, from which originate heavy agricultural products, chiefly wheat, but which are also traversed by the chief transcontinental railroads.

Products of mines made up more than one-half of the freight tonnage in 1946, with coal by far the most important single commodity (more than one-third of the total tons carried).[2] Several of the largest railroad towns are on coal or ore carrying railroads. Duluth-Superior is the terminal of the short iron-ore railroads from the Mesabi Range. Huntington-Ashland is situated in the coal fields of eastern Kentucky, West Virginia and contiguous Virginia, from which originate enormous quantities of coal shipped by railroad to Lake Erie ports— chiefly to Toledo, the world's largest coal port[3]—and to Hampton Roads. Roanoke is

1. Ullman, *op. cit.*, 1949, p. 252.
2. Ullman, *op. cit.* 1949, p. 251.
3. A. G. Ballert, "The Coal Trade of the Great Lakes and the Port of Toledo", *Geographical Review* (1948).
4. Harold M. Mayer, "Railroads and City Planning," *Journal of the American Institute of Planners* (Number 4, 1946), p. 9.

located east of these coal fields on the railroads to Hampton Roads. Scranton is in the anthracite field of northeastern Pennsylvania.

Albany-Troy, at the head of deepwater navigation on the Hudson River and at the great bend of the Mohawk Route, has long been an important transportation center. Its railroad functions are connected with its location on the New York Central Railroad, which follows the water-level route of the New York State Barge Canal (the improved Erie Canal). New York Central has four tracks most of the way between New York and Chicago and it has the next heaviest traffic of American railroads after the Pennsylvania.

Harrisburg has one of the largest classification yards in the United States (Enola). Here are classified cars to and from all points south and east on the one hand, and north and west on the other. For example, a train from Chicago may enter the yard with cars for New York, Philadelphia and Baltimore. These cars are sorted, and all cars for New York are combined with cars for the same destination which arrived at Enola in trains from St. Louis, Cleveland, and so on.[4]

Altoona, the largest American railroad city of A-type, owes its existence to the Pennsylvania Railroad, which maintains huge repair shops in the city.

On Jacksonville, the important railroad center of northern Florida, converge lines from the Northeast and the Midwest, from the fruit and vegetable districts of central Florida and from the resorts along Florida's east coast. The line from Miami to the large cities of the Atlantic seaboard has the heaviest traffic. Savannah is also located on this railroad.

The Southeast, with a rather dense net of railways, is an area of light traffic according to Ullman's map. Atlanta, originating as a railroad center, stands out as the leading transportation focus, with the same percentage for railroad employment as Chicago. As a close parallel with Chicago-Pittsburgh, the steel and mining center of Birmingham has an even higher railway employment percentage (4.6).

Little Rock, Springfield and Topeka have large repair shops, making them important railroad towns. Lincoln has a large classification yard.

Omaha, largest among the cities here classified as railroad towns, is an important transportation center. On Ullman's map, Kansas City (railroad employment percentage 5.1), Minneapolis-St. Paul (4.7) and Omaha (9.1) stand out as leading railroad foci in the western part of the Middle West, second only to Chicago and St. Louis. Council Bluffs, the twin city of Omaha on the Iowa side of the Missouri, in 1863 became the eastern terminus of America's first transcontinental railroad, the Union Pacific. On this line, which has the heaviest freight traffic of the transcontinental railways, is a string of railroad towns among which can be mentioned North Platte, Nebraska, Cheyenne, Wyoming, Ogden, Utah, and Sacramento between Omaha and San Francisco and Pocatello, Idaho, on the branch line to Portland.

San Bernardino, the city on which for topographic reasons the transcontinental railways to Los Angeles converge, Tucson and El Paso constitute another string of railroad towns. On the northern routes Spokane is the largest railroad city.

### TRUCKING AND WAREHOUSING

Highway transportation is far more flexible than the other forms of transportation. Private cars dominate the inter-city passenger movement, but busses are also important. Trucks carry an increasing share

of inter-city freight. The flexibility of highway transportation means a ubiquitous distribution of people commercially engaged in this industry (Fig. 21);[1] highway traffic does not directly create new cities, nor does it substantially influence the growth rate of existing cities. Along the main transcontinental highways in the sparsely populated western parts of the United States there are many small towns, the chief function of

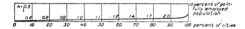

FIG. 21. Cumulative distribution diagram for trucking and warehousing.

which is to provide gasoline for the cars and trucks, and meals and lodging for drivers and passengers. Many of them would not exist except for the highway traffic, yet they are not highway transportation towns but a combination of retail trade, motel and restaurant towns. Their inhabitants are not engaged in highway traffic but in serving the traffic-flow through the town.

### OTHER TRANSPORTATION

"Other transportation" is a highly ubiquitous industry (Fig. 22). Shipping seems to be the chief city forming element, indicated

FIG. 22. Cumulative distribution diagram for other transportation.

by the fact that the few other-transportation towns are all port cities and also by the high percentages for San Francisco, New York and Baltimore in comparison with the other cities of more than a million inhabitants (Appendix 2). Also air transportation

---

1. This industry includes only trucking and warehousing, but it should be representative of commercial highway traffic in general.

employs proportionally more people in some cities than in others. Since the airways are most competitive in long distance passenger traffic, only the largest cities will serve as terminals or intermediate stops for air routes of any considerable importance.[1] Only the largest cities have a sufficient traffic volume between themselves to support such routes. The relatively small effect of air transportation on city growth will thus be to strengthen further the position of the largest cities. New York is the American terminal with most lines on the important trans-Atlantic route. San Francisco has a similar position with regard to the smaller trans-Pacific traffic. Domestic routes, however, dominate the American air transportation pattern.

The only American other-transportation town of B-type is Galveston, Texas. This city has grown rather slowly in recent decades but it is still a large port. The completion of the Houston Ship Canal in 1914 was a serious handicap for Galveston in the competition for the bulky goods shipped from the Gulf Coast of Texas. Galveston is a "dry cargo" port, shipping principally cotton, wheat and sulphur, whereas petroleum products make up more than 80 per cent of Houston's shipments.[2]

The C-type cities are: Miami, Mobile, New Orleans, Astoria, Oregon, Ashtabula, Ohio, and Sault Ste. Marie, Michigan. The port of Miami is chiefly a passenger port for winter tourists going from New York to Florida, Cuba and the Gulf Coast; it handles little freight. Miami is also a notable air transportation center, an important stop on the lines to Latin America. Southern Florida itself generates much traffic since many tourists travel by air. Mobile and New Orleans are important port cities. Mobile exports cotton, forest products and iron and steel products from the Birmingham district and imports bauxite and other products chiefly from Latin America. New Orleans exports petroleum, cotton and lumber, and imports coffee, sugar, bananas and other products mainly from Latin America. Astoria, near the mouth of the Columbia River, is one of many small cities on the coast of the Pacific Northwest which owe their existence to lumber-export activities. Ashtabula is one of the ore ports on Lake Erie; Sault Ste. Marie is located on the largest waterway in the world, the Soo Canals.

## TELECOMMUNICATIONS

Telecommunications constitute one of the most ubiquitous American industries (Fig. 23). No telecommunication towns are found in the United States.

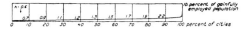

FIG. 23. Cumulative distribution diagram for telecommunications.

## UTILITIES AND SANITARY SERVICES

Utilities and sanitary services also make up a highly ubiquitous industry (Fig. 24). Dixon, Illinois, Fergus Falls, Minnesota, and Florence, Alabama, have a high enough employment percentage for this industry to make them C-type towns. No explanation has been found for this unexpected concentration of employment in the utilities and sanitary services industry, which comprises electric and gas utilities, water supply, sanitary services, and other utilities.

FIG. 24. Cumulative distribution diagram for utilities and sanitary services.

---

1. Not less than 63.3 per cent of the employees of this industry live in cities with more than 250,000 inhabitants as compared to 40.0 per cent for the total gainfully employed population, Fig. 1.

2. J. J. Parsons, "Recent Industrial Development in the Gulf South," *Geographical Review* (1950).

TABLE 6. Correlation Table: Wholesale Trade / Other Retail Trade.
I–X, Wholesale Trade, decils. A–J, Other Retail Trade, decils. For calculation of decils, see page 14 and the diagrams of Figs. 25 and 29.

|   | I | II | III | IV | V | VI | VII | VIII | IX | X | No. of cities |
|---|---|---|---|---|---|---|---|---|---|---|---|
| A | 55 | 17 | 8 | 4 | 2 | — | — | — | — | — | 86 |
| B | 15 | 23 | 15 | 12 | 9 | 7 | 5 | — | 1 | — | 87 |
| C | 5 | 19 | 21 | 12 | 8 | 7 | 4 | 8 | 1 | 1 | 86 |
| D | 7 | 12 | 11 | 12 | 10 | 11 | 13 | 6 | 4 | 1 | 87 |
| E | 2 | 7 | 12 | 12 | 12 | 13 | 8 | 9 | 4 | 7 | 86 |
| F | — | 4 | 12 | 11 | 14 | 11 | 8 | 9 | 13 | 5 | 87 |
| G | — | — | 1 | 9 | 14 | 14 | 10 | 12 | 15 | 11 | 86 |
| H | 2 | 2 | 3 | 5 | 8 | 13 | 16 | 18 | 12 | 8 | 87 |
| I | — | 1 | 1 | 6 | 7 | 6 | 17 | 11 | 19 | 18 | 86 |
| J | — | 1 | 2 | 3 | 2 | 4 | 6 | 14 | 18 | 36 | 86 |
| No. of cities | 86 | 86 | 86 | 86 | 86 | 86 | 87 | 87 | 87 | 87 | 864 |

TABLE 7. Correlation Table: Wholesale Trade/Manufacturing, Including Mining and Construction
I–X, Wholesale Trade, decils. A–J, Manufacturing, decils. For calculation of decils, see page 14 and the diagrams of Map 1 and Fig. 25.

|   | I | II | III | IV | V | VI | VII | VIII | IX | X | No. of cities |
|---|---|---|---|---|---|---|---|---|---|---|---|
| A | 9 | 7 | 9 | 9 | 8 | 9 | 7 | 7 | 8 | 13 | 86 |
| B | — | 2 | 4 | 9 | 8 | 6 | 8 | 9 | 17 | 24 | 87 |
| C | — | 4 | 3 | — | 4 | 8 | 12 | 14 | 23 | 18 | 86 |
| D | 2 | 1 | 3 | 6 | 6 | 8 | 14 | 17 | 14 | 16 | 87 |
| E | — | 4 | 3 | 4 | 12 | 13 | 11 | 14 | 13 | 12 | 86 |
| F | 2 | 4 | 9 | 12 | 15 | 9 | 13 | 13 | 8 | 2 | 87 |
| G | 2 | 6 | 12 | 16 | 11 | 11 | 13 | 9 | 4 | 2 | 86 |
| H | 7 | 10 | 16 | 10 | 14 | 18 | 9 | 3 | — | — | 87 |
| I | 16 | 23 | 21 | 14 | 8 | 3 | — | 1 | — | — | 86 |
| J | 48 | 25 | 6 | 6 | — | 1 | — | — | — | — | 86 |
| No. of cities | 86 | 86 | 86 | 86 | 86 | 86 | 87 | 87 | 87 | 87 | 864 |

A–I: Princeton, N.J.; Kings Park, N.Y.; State College, Pa.; Stillwater, Okla.; Laramie, Wyo.; Pullman, Wash.; Ames, Iowa; Midway-Hardwick, Ga.; and Auburn, Ala.

A–II: Annapolis, Md.; El Reno, Okla.; Boulder, Col.; Las Vegas, Nev.; Oceanside, Calif.; Corvallis, Ore.; and Iowa City, Iowa.

A–III: Newport, R.I.; Washington, D.C.; Norman, Okla.; Lawrence, Kan.; Junction City, Kan.; Tallahassee, Fla.; St. Augustine, Fla.; Gainesville, Fla.; and Key West, Fla.

B–II: Hornell, N.Y.; and Ann Arbor, Mich.

## Trade Industries

### WHOLESALE TRADE

Wholesale trade is a ubiquitous industry in American cities, but its importance varies considerably from town to town. Cities in the tenth decil have more than five times as many people employed in wholesale trade as those of the same size in the first decil.

There is a good correlation between employment in wholesale trade and in retail trade, indicating an association of

TABLE 8. Regional Differentiation of Wholesale Trade as a City Forming Activity.
I–X, Wholesale Trade, decils. A map of the eight zones is on page 85.

| Zones | I | II | III | IV | V | VI | VII | VIII | IX | X | No. of cities |
|---|---|---|---|---|---|---|---|---|---|---|---|
| Northeast . . . . . . . . | 28 | 19 | 16 | 14 | 13 | 8 | 4 | 2 | 1 | 3 | 108 |
| East Manufacturing . . . . | 16 | 20 | 15 | 12 | 13 | 11 | 13 | 6 | 2 | 0 | 108 |
| West Manufacturing. . . . | 9 | 16 | 16 | 23 | 10 | 11 | 9 | 11 | 3 | 0 | 108 |
| North-Central . . . . . . | 8 | 7 | 6 | 7 | 13 | 15 | 12 | 19 | 12 | 9 | 108 |
| Southeast . . . . . . . . | 11 | 10 | 8 | 11 | 8 | 12 | 15 | 17 | 11 | 5 | 108 |
| South . . . . . . . . . . | 5 | 4 | 10 | 9 | 10 | 10 | 12 | 13 | 17 | 18 | 108 |
| Prairie. . . . . . . . . . | 2 | 4 | 4 | 7 | 12 | 7 | 12 | 10 | 22 | 28 | 108 |
| West . . . . . . . . . . | 7 | 6 | 11 | 3 | 7 | 12 | 10 | 9 | 19 | 24 | 108 |
| No. of cities . . . . . . . | 86 | 86 | 86 | 86 | 86 | 86 | 87 | 87 | 87 | 87 | 864 |

TABLE 9. Correlation Table: Wholesale Trade/City Size.
I–X, Wholesale Trade, decils.

| City size 1,000 inh. | I | II | III | IV | V | VI | VII | VIII | IX | X | No. of cities |
|---|---|---|---|---|---|---|---|---|---|---|---|
| 10–25 | 70 | 57 | 55 | 44 | 55 | 45 | 51 | 39 | 43 | 48 | 507 |
| 25–50 | 12 | 19 | 19 | 19 | 19 | 16 | 16 | 19 | 14 | 18 | 171 |
| 50–100 | 3 | 4 | 7 | 6 | 5 | 9 | 6 | 8 | 2 | 10 | 60 |
| 100–250 | 1 | 6 | 4 | 12 | 4 | 8 | 10 | 10 | 15 | 7 | 77 |
| 250–500* | — | — | — | 4 | 2 | 3 | 2 | 6 | 5 | 2 | 24 |
| 500–1,000* | — | — | — | — | 1 | 2 | — | 2 | 6 | 2 | 13 |
| 1,000–* | — | — | 1 | 1 | — | 3 | 2 | 3 | 2 | — | 12 |
| No. of cities | 86 | 86 | 86 | 86 | 86 | 86 | 87 | 87 | 87 | 87 | 864 |

* *Cities with more than one million inhabitants: 9th decil:* San Francisco, New York; *8th decil:* Los Angeles, Boston, St. Louis; *7th decil:* Baltimore, Chicago; *6th decil:* Cleveland, Pittsburgh, Philadelphia; *4th decil:* Detroit; *3rd decil:* Washington.
   *Cities with 500,000–1,000,000 inhabitants: 10th decil:* Portland, Dallas; *9th decil:* Kansas City, Minneapolis-St. Paul, Houston, Atlanta, Seattle, New Orleans; *8th decil:* Indianapolis, Cincinnati; *6th decil:* Buffalo, Milwaukee; *5th decil:* Providence.
   *Cities with 250,000–500,000 inhabitants: 10th decil:* Memphis, Omaha; *9th decil:* San Antonio, Oklahoma City, Richmond, Denver, Fort Worth; *8th decil:* Miami, Norfolk-Portsmouth, Nashville, Columbus, Louisville, Birmingham; *7th decil:* Albany-Troy, Syracuse; *6th decil:* Toledo, Hartford, San Diego; *5th decil:* Rochester, Springfield-Holyoke; *4th decil:* Dayton, Youngstown, Akron, Wilkes-Barre.

typological significance (Table 6). There is an inverse correlation between employment in wholesale trade and in manufacturing (Table 7). A city is likely to have a low wholesale trade percentage if it has a high manufacturing rate, and vice versa. Cities in the lowest manufacturing decil show an undecided tendency. Most one-sided service cities, such as university towns, hospital towns and so on, belong to this decil. They are often as "underdeveloped" in some of the other service industries as one-sided manufacturing cities.

More than two-thirds of the cities in the Western and Prairie zones have an above-average percentage for employment in wholesale trade (Table 8). In the Northeast zone, the East Manufacturing zone and the

West Manufacturing zone the same proportion of cities are below average. The regional differentiation is, however, not so fully pronounced for wholesale trade as for retail trade.

Wholesale trade is well developed in cities with more than 100,000 inhabitants, most of which have a greater-than-average percentage for wholesale trade employment (Table 9). For retail trade the opposite is true: most big cities have a rate below average (Table 10). The difference in this respect between wholesale and retail trade reflects the vertical structure of trade. Retail stores serve the ultimate consumer and are located accordingly. There is a definite limit to the area which can conveniently be served by a retail establishment. Even if this limit is different in various lines of business, the corresponding limit for a wholesale establishment serving a large number of retail stores will always be wider. From this it follows that: (1) The retail trade area of a big city, even if it is bigger than that of a small one, cannot be expected to be proportionally bigger; thus retail trade will be of smaller relative importance in the big city than in the small one. (2) In wholesale trade the big city will often appear as the natural location, because such a city in itself offers a large market and because it is the hub of an intense transportation net by which the trade area of the wholesale establishment can efficiently be served.

New York and San Francisco have the highest wholesale trade percentage among the metropolises with more than one million inhabitants. They are followed by Los Angeles, Boston and St. Louis. In this size-group Washington and Detroit, the only one-sided cities with more than a million inhabitants, are the only ones that fall below the median average for cities with more than 10,000 inhabitants.

Fig. 25, showing cities with a wholesale trade "surplus" of 5 per cent or more, reveals an interesting pattern. Cities of this type are almost missing in the Manufacturing Belt, with about half of all cities in the United States. There are some exceptions in the diffuse border zone of this area: Green Bay, Wisc.; Burlington, Iowa; Gloucester, Mass.; and Portland, Me. As can be seen from Table 8, all decils except the two highest are well represented in the Manufacturing Belt; this means that there is a considerable variation between cities as centers of wholesale trade even in this region with its high city density.

The largest concentration of strongly marked wholesale cities is found in a wide belt with north-southerly extension, stretching through the Middle West and the South and indicated by the following cities: Fargo, Duluth, Minneapolis - St. Paul, Sioux Falls, Sioux City, Omaha, St. Joseph, Independence, Memphis, Little Rocks, Fort Smith, Dallas, Waco, New Orleans. A typical town of this area has approximately as many people employed in wholesale trade as a New England city with three times as many inhabitants. All cities with more than 250,000 inhabitants in this belt fall within the ninth and tenth wholesale decils. St. Louis, in the border zone between this area and the Manufacturing Belt, belongs to the eighth decil.

Other areas in which wholesale trade frequently is an important city forming industry are Central Florida, the lower Rio Grande Valley, the Panhandle of Texas, and the states west of the Rocky Mountains.

The very highest wholesale percentages are found in small cities serving areas of intensive agriculture producing fruits and vegetables.[1] Many people are here employed

1. All B-centers are of this type: Corona (lemons), Santa Paula (citrus), and Salinas (vegetables) in California; Mercedes (citrus and vegetables), San Benito (citrus), and Edinburg (citrus) in the lower Rio Grande Valley; Sanford (celery), and Fort Pierce (citrus) in central Florida; and Suffolk (peanuts) in Virginia.

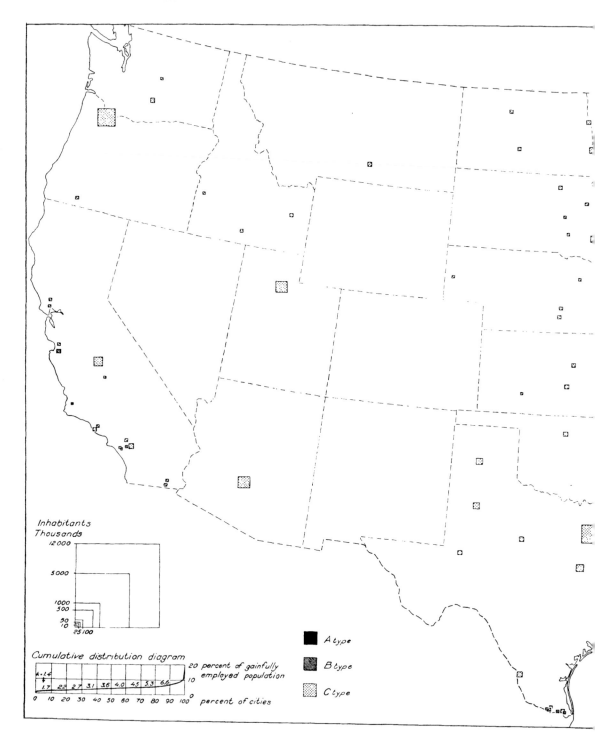

Inhabitants
Thousands

Cumulative distribution diagram

20 percent of gainfully employed population

percent of cities

A type
B type
C type

FIG. 25. Wholesale trade towns are conspicuously concentrated in a wide belt in central United States stretching from the Canadian border to the Gulf Coast and the border of Mexico. They are as conspicuously missing in the Manufacturing Belt. Wholesale trade is also an important city forming activity in the West and partly also in the South.

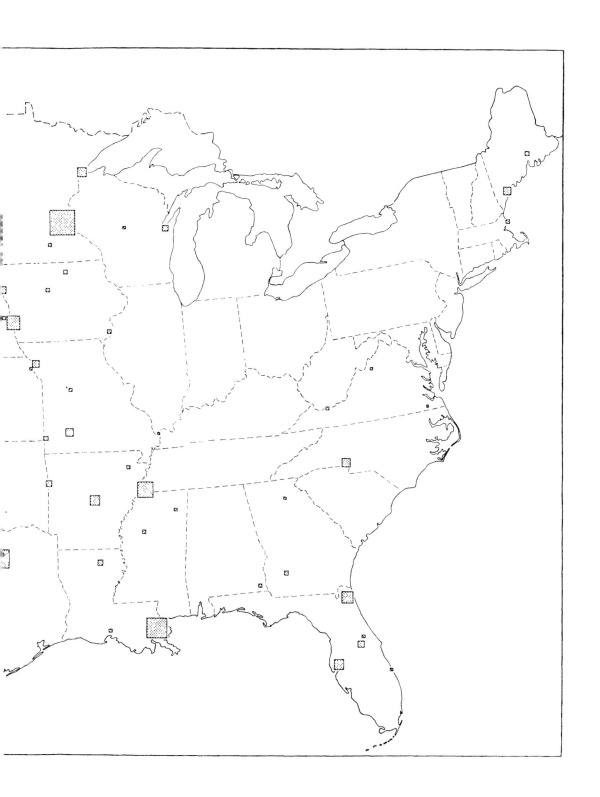

in packing houses, grading and packing the products for the national and international market.

Space does not permit a discussion of the importance of wholesale trade in individual centers. Only a few general remarks shall be made about some of the biggest cities.

*Large Port Cities of the Atlantic Seaboard*

There seems to be an unanimous opinion among economic geographers and historians that New York's position as the biggest city on the American continent is chiefly a function of its location on the Hudson River. The opening of the Erie Canal in 1825, connecting Lake Erie with the Hudson River, lowered freight rates between Buffalo and New York to 5 per cent of what they had been before.[1] New York, already the biggest city and port in the United States, became the undisputed leader among American cities. The Hudson River became the natural outlet of the Great Lakes region. The Erie Canal made possible the settlement of the Middle West, which was to become the dominant American farm region.

When the railroads, a few decades later, surpassed the Erie Canal as carriers of traffic between the Middle West and the Atlantic Seaboard, New York's trade monopoly was broken. The city had, however, gained a position during the heyday of the Erie Canal (1825 to approximately the 1870's) that Philadelphia, Baltimore or Boston could have challenged only if the railroads had given them definite advantages over New York.

New York early got a foreign trade exceeding that of all other Atlantic ports

combined. Goods and passengers from all parts of the world converged in its big port. The city became the chief American terminal on the dominating trade route of the world, connecting northeastern United States with northwestern Europe. The localization of such important New York industries as apparel manufacturing, banking, wholesale trade and others can be shown to have been strongly influenced by New York's position as the chief port of entry for European trade, European immigrants and European influences in general.

Boston is closer to northwest Europe than New York, Philadelphia or Baltimore, but its hinterland is confined to New England by its great distance from the Middle West, which gives advantages to the three competing metropolises. Thus Boston has a small but populous hinterland. In its ocean-borne trade incoming goods greatly exceed outgoing goods, leading to unfavorable freight rates. Boston, among other things, has the largest wool market in the United States.

Baltimore, on a navigable river near the Inner Chesapeake Bay, is the most southerly of the four leading northeastern port cities. It is nearer to the Middle West than its rivals, but farther from northwest Europe. Baltimore early became an important center for shipping and trade. The more recent development of important manufacturing industries in the city can be seen as a result of the favorable trade location.

Philadelphia, with the second largest American port, is proportionally the most industrialized of the four leading cities on the Atlantic Seaboard and it also has the lowest wholesale trade rate. About the middle of the eighteenth century Philadelphia surpassed both New York and Boston in trade and population, chiefly because it had a larger local hinterland than the other cities. Manufacturing in Philadelphia is highly diversified. Some of the most important industries

1. White and Foscue, *Regional Geography of Anglo-America* (New York, 1954), p. 68. Hicks (*The Federal Union*, New York 1952, p. 357) states that the canal reduced the cost to 10 per cent, and Faulkner (*American Economic History*, New York 1943, p. 279) mentions that the cost dropped to 15 per cent. All these authors agree that the transit time was reduced from 20 to 8 days.

have been attracted by the advantageous port location: oil-refining, sugar refining, fertilizer manufacturing, ship-building, and (after 1950) steel manufacturing based on imported ore.

### Chicago and St. Louis

In the Middle West, Chicago and St. Louis are the leading trade centers. They are located in the center of one of the world's largest areas of intensive agriculture,[1] which in its eastern part is superimposed by the western section of the world's leading manufacturing belt.

St. Louis was founded in 1764 as a fur-trading post. In the early part of last century it became the leading outfitting center for farmers settling the prairies and miners moving into the mountains of the West. Through the development of numerous overland trails, St. Louis acquired a very wide trade area. Its strategic river location enabled the city to dominate the steamboat traffic on the Mississippi. During the latter half of the last century St. Louis became a railroad center, ranking second only to Chicago. St. Louis' location was favorable for many manufacturing activities and the city developed into one of the most diversified manufacturing centers in the United States.

Chicago, located at the point where the Great Lakes and Mississippi Valley waterways meet, rapidly expanded with the Middle West to become the natural economic capital of this great agricultural area and the major transportation focus of central North America. The city overtook St. Louis in population about 1870. The early railroads were feeding lines to the Great Lakes, which through the Erie Canal had an outlet to the world market via New York. When railroads were built from the eastern seaboard cities, they found it necessary to

reach Chicago to connect with the rapidly expanding western railroad systems.[2]

Chicago's position as the economic capital of the agricultural Middle West is reflected by one of the world's leading grain exchanges and by the largest livestock market in the United States. Like New York it is also the headquarters of many nationally known firms and it is a leading center for mail-order houses.

Wholesale trade, railroads and food manufacturing, important as they have been for the growth of Chicago, have, however, been even more important for St. Louis, as can be seen from Appendix 2.[3] Chicago's industrial structure also reveals many similarities with Cleveland and other big cities on the Lower Lakes. Metal manufacturing, ranging from basic steel to different types of machinery, is by far the leading group of cityforming industries. In this respect Chicago differs from other big cities outside of the Lower Lakes—Pittsburgh area.

### Large Cities in Central United States

All big cities in central United States, near the north-southerly, climatically determined division zone between intensive agriculture to the east and extensive grazing and wheat farming to the west,[4] have a high wholesale percentage: Minneapolis - St. Paul, Omaha, Kansas City, Oklahoma City, Dallas, Fort Worth, and San Antonio. They are located in the periphery of the trade areas of Chicago and St. Louis. Their industrial structures

1. The Agricultural Interior Region according to White and Foscue, *op. cit.*, p. 6–7.
2. H. M. Mayer and A. K. Philbrick, "Chicago". *Industrial Cities Excursion, Guidebook.* XVIIth International Geographical Congress (Washington, 1952), p. 42.
3. Wholesale trade and food manufacturing appear, besides retail trade, with the greatest frequency as leading city-forming industries in the agricultural areas west of the Manufacturing Belt as well as in other predominantly agricultural areas (Appendix 1).
4. For a map of the chief geographic regions of Anglo-America, see White and Foscue, *op. cit.*, p. 6.

are similar, strongly influenced by their functions as contact points between two physically contrasting but economically interrelated regions. Cattle from the western ranges are fattened on farms surrounding these cities, sold in their stockyards, slaughtered in their packing-houses and sent in the form of meat to the eastern market. Wheat is stored or milled in these cities on its way from the western dry-lands to the eastern market. Industrial products from the Manufacturing Belt are distributed to farmers and ranchers in the wide trade areas of these cities.

Minneapolis-St. Paul (the Twin Cities) is the largest urban center in a wide area between the Great Lakes and the Rocky Mountains. Minneapolis, located at waterfalls in the Mississippi, early became a leading milling center for the Spring Wheat Region. Like many cities in the Great Lakes area, it was also an important sawmill town during the period of forest exploitation. St. Paul, a few miles below, was at the head of navigation, a strategic location in the pre-railroad days. The economic activity of the large trade area of the Twin Cities is reflected in the big flour mills of Minneapolis and the stockyards and packing plants of St. Paul. The diversified industrial structure of the Twin Cities, with a rather low manufacturing percentage, shows larger similarities with those of Kansas City and Omaha than with those of big cities in the Manufacturing Belt.

Omaha and Kansas City are the biggest of four cities on the Missouri River situated near the western margin of the Corn Belt and close to the grazing lands and the Winter Wheat Belt farther west. The others are Sioux City and St. Joseph. They are leading cattle markets with large stockyards and packing plants, having benefitted from the westward movement of the packing industry. Kansas City is America's largest

market for winter wheat and one of the leading milling centers.

Oklahoma City, situated on the border between the Winter Wheat Belt and the Cotton Belt, grew up as a commercial center for an agricultural area producing chiefly cattle, wheat and cotton. The trade area of Oklahoma City has subsequently been enriched by the exploitation of the huge Mid-Continent Oil Field. This development is also characteristic of the two leading centers of northern Texas, Dallas and Fort Worth. Their downtown areas are only about 30 miles apart, an unusual location for large commercial cities in the inland. There is, however, a rather distinct specialization in their trade functions. Dallas is chiefly a distributing center in which are located the branch offices for the state or the Southwest of most firms working on the national market. The importance of Dallas as a financial center was underlined when it was chosen as the location for one of the twelve Federal Reserve Banks. That Fort Worth is a market center for the range country to the west is reflected in its large stockyards and packing plants. Both cities also have flour mills and cottonseed-oil mills as well as oil refineries and factories making oil-field equipment. Dallas and Fort Worth are located in the northern part of the Black Prairie of Texas, a fertile and relatively densely populated area stretching about 300 miles from north to south.

San Antonio is the trade center of a wide area in southwestern Texas including the range land of the Edwards Plateau to the west, the fruit and vegetable district of the lower Rio Grande Valley to the south, and the Cotton Belt to the east. The chief city forming activities of San Antonio are, however, four huge military airfields, among the largest in the country, and an army post.

Denver, at the foot of the Rocky Mountains, functionally shows similarities with the north-southerly string of big cities from Minneapolis-St. Paul to San Antonio. Like them, it is an economic point of contact between two contrasting types of land: the dry range land of the Great Plains region and the large irrigated area in the South Platte Valley north of the city. As in other irrigated districts the two types of land are intimately related in use: most of the irrigated land is taken up by forage crops for livestock, a large part of which spends some time on the adjacent range. In the economic structure of Denver these agricultural relations are reflected by large stockyards and packing plants.

The size and industrial structure of Denver indicate that it has become the commercial center of a wide area in the sparsely populated Rocky Mountain states. It is also the major financial and administrative center between the Missouri River and the Pacific Coast. The chief office of the Bureau of Reclamation, the federal agency in charge of the big irrigation projects of the West, is located in Denver. The early growth of the city was, however, connected with rich finds of gold and silver in the vicinity. It is still the headquarters of many mining operations in the nearby Rocky Mountains. Denver was not on any of the transcontinental railroads, which preferred easier routes north and south of the city, but it has attracted several railways. In recent years it has become one of the biggest centers for transcontinental trucking.[1]

### Large Port Cities of the Gulf Coast

New Orleans, located about 100 miles from the mouth of the Mississippi, has for two centuries been a leading port city on the North American continent. The city had its greatest relative importance in the early nineteenth century after the Louisiana Purchase, when the Mississippi became an American river, the only water outlet of the rapidly expanding Middle West. The completion of the Erie Canal in 1825 and the construction of several railroads from the interior to the Atlantic Coast a few decades later were a serious threat to the river traffic and the trade of New Orleans. The Mississippi had the great disadvantage of flowing in the wrong direction with regard to the dominant east-westerly trade.

Modern New Orleans is still chiefly a port city of the general cargo type with a good balance between inbound and outbound traffic. The city has a major cotton market, and cotton ranks with petroleum and lumber as the chief outbound commodities. In the trade with Latin America the port ranks second only to New York, as is reflected in the list of inbound goods, dominated by coffee, sugar, bananas and bauxite. The relatively unimportant manufacturing industries of New Orleans are dominated by such typical port-city activities as sugar refining and petroleum refining.

The other big port-city on the Gulf Coast, Houston, is one of the fastest growing cities in the United States. Prior to the completion of the Houston Ship Canal in 1915 Galveston was the leading export port for the western part of the Cotton Belt. With a deep waterway to the Gulf, Houston had a much more favorable location. Its bulky export commodities, petroleum products, cotton and wheat, frequently make it second only to New York in the amount of tonnage handled. Houston is the chief commercial center of the expanding oil fields along the Gulf Coast. Its large oil refineries receive oil from the local fields and from the Mid-Continent Field. In recent years Houston has developed into an important center of

1. Clark N. Crain, "Denver, Colorado, to Salt Lake City, Utah." *Transcontinental Excursion, Guidebook.* XVIIth International Geographical Congress (Washington, 1952), p. 41.

chemical manufacturing, based on local natural resources—petroleum, sulphur, and others.

## Other Large Cities of the South

Richmond is the southernmost of the big cities on the Fall Line, the boundary of the Coastal Plain and the Piedmont. It is accessible for average-sized ocean-going vessels. This location at the head of navigation, with water power available for manufacturing, was very favorable for the early growth of cities in the eastern United States, as is evidenced by the string of big cities further north on the Fall Line: Washington, Baltimore, Wilmington, Philadelphia, Trenton and New York (Paterson). Because of its age and its location at the edge of the tobacco areas of the Piedmont, Richmond has long been a leading tobacco market. Large tobacco factories are located in the city.

Atlanta grew with the railroads from an unimportant town to become the Chicago or St. Louis of the South. Because of its favorable location near the southern end of the Appalachian Mountains, Atlanta was made the meeting point of railroads from the Northeast and the Middle West to the South. It is the chief distributing and financial center of the Southeast.

Memphis is a chief Mississippi port, the largest one between St. Louis and New Orleans. It is the commercial center of a wide region. Memphis' location in the middle part of the Cotton Belt and in the center of the intensive cotton farming areas of the Mississippi Bottoms early made it a leading cotton market. The Mississippi Bottoms region is also the leading hardwood lumbering district in the United States, and Memphis is the largest hardwood market.

## Large Port Cities of the Pacific Coast

San Francisco for a long time held a similar position in the West and in the trade with the Pacific countries as New York held in the East and in the trade with Europe. It got a flying start with the gold rush of 1848, but its natural advantages as a port city were great. It has one of the very few natural ports on the California coast, one of the best on the North American continent. Its immediate hinterland, the Central Valley, is the largest agricultural district of the Pacific coast region. San Francisco's position as the leading wholesale trade and financial center of the West has, however, been challenged in the last decades by Los Angeles. This city, the fastest expanding metropolis in the world, grew up around a center a few miles inland. Citrus growing from about 1885, oil production from the turn of the century, the moving-picture industry from about 1910, catering for retired people seeking a pleasant climate, and later the manufacturing of airplanes were the chief cityforming activities of Los Angeles. Its big port was built after the opening of the Panama Canal in 1914. The port is thus more a result of the big city's ambition to establish itself as a center of waterborne traffic than an original cityforming factor. Both San Francisco and Los Angeles have been terminals on transcontinental railroads since the end of last century.

Seattle and Portland, the two leading centers of the Pacific Northwest, have grown in response to favorable trade locations. Portland, with one of the highest wholesale trade rates among big cities in the United States, has a natural crossroads position at the junction of the main north-south route from California to Puget Sound and the only water-level passage from the Columbia Basin to the coast. The fertile Willamette Valley lies within its immediate hinterland.

Of the Puget Sound cities, Seattle and Tacoma were nearly equal in population before 1900, but Seattle forged ahead thanks to better railway and ocean shipping facilities. Of great importance for the growth of the city was the Klondike gold rush in 1898. Seattle became the recognized gateway to Alaska, a position which the city still retains. In steaming distance Seattle is nearly 2 days closer to the Orient than either of its rivals, San Francisco or Los Angeles. In the 1920's the city received a lion's share of the raw silk import from Japan. This valuable commodity was trans-shipped by fast trains to the New York region. In 1929 raw silk accounted for 69 per cent, or about 150 million dollars, of the total import value over the Washington Customs district.[1]

Seattle and Portland are like San Francisco at first-hand centers of trade, finance and administration. They are not so industrialized as Los Angeles and the big cities in the Northeast. Both foreign and domestic outbound traffic, the latter by far the more important, is dominated by forest products and agricultural commodities, chiefly wheat and fruit. The chief foreign trade partners for Seattle and Portland, as well as for San Francisco and Los Angeles, are the Asiatic countries.[2]

## FOOD STORES

Sale at retail of food and dairy products, including milk, constitute a large American industry, employing about twice as many people as the manufacturing of automobiles (Fig. 1). But since food stores are highly ubiquitous (Fig. 26), there are no food store towns in the United States.

FIG. 26. Cumulative distribution diagram for food stores.

## EATING AND DRINKING PLACES

Eating and drinking places are ubiquitous in American cities (Fig. 27). In twelve towns, all of C-type, this industry ranks as a major cityforming activity. Most of these are recognized as tourist centers: Atlantic City, Daytona Beach and Key West, Florida, Hot Springs, Arkansas, Las Vegas,

FIG. 27. Cumulative distribution diagram for eating and drinking places.

Nevada, Oceanside, Newport Beach and Monterey, California. One is a college town: Bowling Green, Ohio. The remaining ones, Junction City, Kansas, and Merced and Watsonville, California, are retail trade centers. A higher-than-average percentage for employment in eating and drinking places is often found for tourist centers, college towns, medical centers, and retail trade centers. High rates for hotels and eating and drinking places are the best indicators of a tourist center in the industrial structure of American cities.

## OTHER RETAIL TRADE

Other retail trade, employing 5.2 million people, is by far the largest of 36 specified American urban industries (Fig. 1). Retail trade as a whole, including also food stores and eating and drinking places, employs more people than agriculture (8.2 million and 6.9 million respectively).

All cities with more than one million inhabitants and the majority of those of 250,000 or more have a below median average percentage for employment in other

1. O. W. Freeman and H. H. Martin, *The Pacific Northwest* (New York, 1954), p. 430.
2. "Waterborne trade of the Pacific Northwest." *Monthly Review* (August, 1952). Federal Reserve Bank of San Francisco.

TABLE 10. Correlation Table: Other Retail Trade/City Size.
  I–X, Other retail trade, decils. For the calculation of decils, see page 14 and the diagram of Fig. 29.

| City size 1,000 inh. | I | II | III | IV | V | VI | VII | VIII | IX | X | No. of cities |
|---|---|---|---|---|---|---|---|---|---|---|---|
| 10–25 | 57 | 31 | 46 | 43 | 46 | 50 | 50 | 58 | 55 | 71 | 507 |
| 25–50 | 15 | 17 | 17 | 15 | 26 | 18 | 14 | 20 | 20 | 9 | 171 |
| 50–100 | 7 | 8 | 8 | 10 | 4 | 5 | 4 | 3 | 7 | 4 | 60 |
| 100–250 | 7 | 16 | 5 | 13 | 6 | 9 | 12 | 3 | 4 | 2 | 77 |
| 250–500* | 0 | 6 | 5 | 3 | 2 | 1 | 5 | 2 | 0 | 0 | 24 |
| 500–1,000* | 0 | 3 | 1 | 2 | 1 | 4 | 1 | 1 | 0 | 0 | 13 |
| 1,000–* | 0 | 6 | 4 | 1 | 1 | 0 | 0 | 0 | 0 | 0 | 12 |
| No. of cities | 86 | 87 | 86 | 87 | 86 | 87 | 86 | 87 | 86 | 86 | 864 |

* *Cities with more than one million inhabitants: 5th decil:* Los Angeles; *4th decil:* San Francisco; *3rd decil:* Boston, Baltimore, St. Louis, Philadelphia; *2nd decil:* Chicago, Pittsburgh, Washington, Cleveland, New York, Detroit.
  *Cities with 500,000–1,000,000 inhabitants: 8th decil:* Dallas; *7th decil:* Kansas City; *6th decil:* Atlanta, Portland, Seattle, Minneapolis-St. Paul; *5th decil:* Indianapolis; *4th decil:* Houston, New Orleans; *3rd decil:* Cincinnati; *2nd decil:* Providence, Milwaukee, Buffalo.
  *Cities with 250,000–500,000 inhabitants: 8th decil:* San Antonio, Oklahoma City; *7th decil:* Miami, San Diego, Memphis, Fort Worth, Denver; *6th decil:* Richmond; *5th decil:* Norfolk-Portsmouth, Omaha; *4th decil:* Nashville, Columbus, Hartford; *3rd decil:* Birmingham, Syracuse, Louisville, Toledo, Rochester; *2nd decil:* Albany-Troy, Springfield-Holyoke, Akron, Youngstown, Dayton, Wilkes-Barre.

TABLE 11. Correlation Table: Other Retail Trade/Manufacturing, Including Mining and Construction
  I–X, Other retail trade, decils. A–J, Manufacturing, decils. For calculation of decils, see page 14, Map 1 and Fig. 29.

| | I | II | III | IV | V | VI | VII | VIII | IX | X | No. of cities |
|---|---|---|---|---|---|---|---|---|---|---|---|---|
| A | 6 | 2 | 3 | 7 | 9 | 10 | 3 | 10 | 12 | 24 | 86 |
| B | 0 | 2 | 3 | 3 | 5 | 5 | 13 | 14 | 20 | 22 | 87 |
| C | 0 | 0 | 2 | 5 | 6 | 3 | 10 | 20 | 20 | 20 | 86 |
| D | 1 | 4 | 2 | 8 | 8 | 10 | 13 | 15 | 14 | 12 | 87 |
| E | 0 | 2 | 3 | 9 | 7 | 14 | 19 | 9 | 15 | 8 | 86 |
| F | 2 | 3 | 11 | 8 | 9 | 19 | 19 | 13 | 3 | 0 | 87 |
| G | 1 | 6 | 16 | 16 | 21 | 14 | 7 | 3 | 2 | 0 | 86 |
| H | 2 | 17 | 18 | 19 | 17 | 9 | 2 | 3 | 0 | 0 | 87 |
| I | 7 | 36 | 26 | 10 | 4 | 3 | 0 | 0 | 0 | 0 | 86 |
| J | 67 | 15 | 2 | 2 | 0 | 0 | 0 | 0 | 0 | 0 | 86 |
| No. of cities | 86 | 87 | 86 | 87 | 86 | 87 | 86 | 87 | 86 | 86 | 864 |

retail trade (Table 10). As was previously mentioned (page 98), the relatively low retail trade percentage in the largest cities can be explained by a smaller importance of the outside trade area for the large cities than for the small ones or those of moderate size. The average New York or Chicago store should have relatively fewer visits by out-of-town customers than stores in most other American cities. Considered abso-

TABLE 12. Regional Differentiation of Other Retail Trade as a City Forming Activity. I–X, Other retail trade, decils. A map of the eight zones is on page 85.

| Zones | I | II | III | IV | V | VI | VII | VIII | IX | X | No. of cities |
|---|---|---|---|---|---|---|---|---|---|---|---|
| Northeast | 43 | 26 | 15 | 9 | 2 | 7 | 3 | 1 | 2 | 0 | 108 |
| East Manufacturing | 16 | 23 | 12 | 20 | 11 | 8 | 6 | 7 | 3 | 2 | 108 |
| West Manufacturing | 6 | 16 | 23 | 16 | 15 | 15 | 8 | 4 | 4 | 1 | 108 |
| North–Central | 6 | 9 | 9 | 13 | 17 | 13 | 12 | 13 | 8 | 8 | 108 |
| Southeast | 11 | 7 | 12 | 11 | 14 | 15 | 14 | 11 | 10 | 3 | 108 |
| South | 0 | 4 | 5 | 9 | 13 | 6 | 25 | 20 | 10 | 16 | 108 |
| Prairie | 0 | 1 | 2 | 3 | 5 | 8 | 10 | 16 | 28 | 35 | 108 |
| West | 4 | 1 | 8 | 6 | 9 | 15 | 8 | 15 | 21 | 21 | 108 |
| No. of cities | 86 | 87 | 86 | 87 | 86 | 87 | 86 | 87 | 86 | 86 | 864 |

lutely, New York, Chicago and the other large cities are, of course, leading retail trade centers in the sense that they have the largest total "surplus" of retail trade employment. But their large out-of-town trade becomes small when it is seen in relation to their size. Retail trade does not rank as a leading cityforming activity in large American cities.[1]

There is a strong negative correlation between Other Retail Trade percentages and Manufacturing rates (Table 11). With a high Manufacturing percentage there is a low retail rate, and vice versa. More than three-fourths of the cities in the tenth manufacturing decil belong to the first retail trade decil. All of these cities fall within the four lowest retail decils. The first manufacturing decil is partly an exception to the above-mentioned inverse correlation. Many one-sided service towns (university towns, hospital towns, etc.) with a low manufacturing percentage also have a low retail rate.[2]

The regional differentiation of retail trade percentages is conspicuous (Table 12). The Prairie zone and the Northeast zone mark the two contrasts. About 90 per cent of the prairie towns have an above-average retail trade percentage. The same proportion of

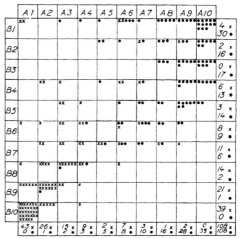

x Northeast  
• Prairie  

A1–A10 decils, other retail trade  
B1–B10 decils, manufacturing (incl. mining and construction)

FIG. 28. Diagram correlating manufacturing percentages and other retail trade percentages for cities of the Northeast zone and the Prairie zone.

cities in the Northeast zone have a below-average rate. A typical city of the Prairie zone belongs to the tenth decil and has about

1. The city forming portion of other retail trade in the large cities may be even smaller than the surplus of the observed retail percentage (for New York 9.2) over the *k*-value (8.0). See pages 18 ff.

2. Examples: *A–I*, Kings Park, N.Y., Midway-Hardwick, Ga., Auburn, Ala., Pullman, Wash., State College, Pa., Princeton, N.J. *A–II*, Ames, Iowa, Washington, D.C.

The same cities also appear as exceptions in the corresponding table for wholesale trade (page 96).

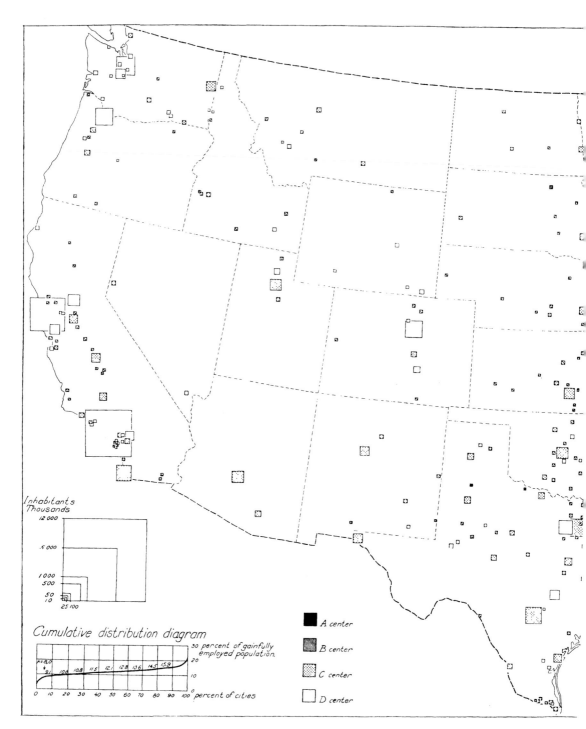

FIG. 29. Other retail trade towns, usually of C-type but some also of B-type, are very common in a wide belt stretching from the Canadian border to the Gulf Coast and the border of Mexico. They are also common in the West but very few are found in the Manufacturing Belt. On this map all cities in the United States are shown.

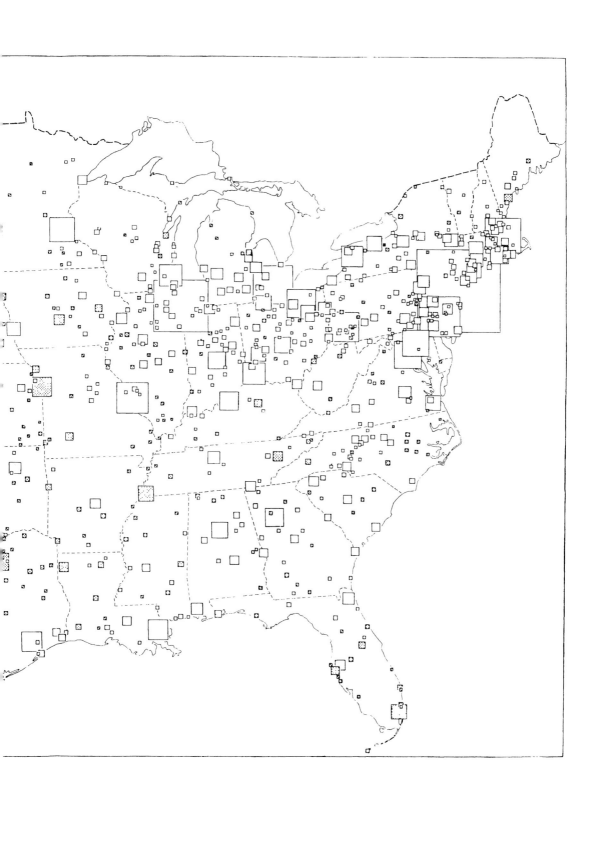

TABLE 13. Correlation Table: Finance, Insurance and Real Estate/City Size. I–X, FIR-decils.

| City size 1,000 inh. | I | II | III | IV | V | VI | VII | VIII | IX | X | No. of cities |
|---|---|---|---|---|---|---|---|---|---|---|---|
| 10–25 | 64 | 61 | 58 | 57 | 54 | 57 | 53 | 40 | 39 | 24 | 507 |
| 25–50 | 15 | 15 | 17 | 19 | 19 | 20 | 19 | 14 | 16 | 17 | 171 |
| 50–100 | 4 | 5 | 5 | 4 | 8 | 5 | 6 | 10 | 5 | 8 | 60 |
| 100–250 | 3 | 5 | 5 | 5 | 5 | 2 | 6 | 17 | 15 | 14 | 77 |
| 250–500* | 0 | 1 | 1 | 2 | 0 | 3 | 0 | 2 | 7 | 8 | 24 |
| 500–1,000* | 0 | 0 | 0 | 0 | 0 | 0 | 2 | 1 | 2 | 8 | 13 |
| 1,000–* | 0 | 0 | 0 | 0 | 0 | 0 | 1 | 2 | 2 | 7 | 12 |
| No. of cities | 86 | 87 | 86 | 87 | 86 | 87 | 87 | 86 | 86 | 86 | 864 |

* Cities with more than 1,000,000 inhabitants: 10th decil: New York, San Francisco, Boston, Los Angeles, Chicago, Washington, Philadelphia; 9th decil: St. Louis, and Baltimore; 8th decil: Pittsburgh, and Cleveland; 7th decil: Detroit.
  Cities with 500,000–1,000,000 inhabitants: 10th decil: Dallas, Seattle, Atlanta, Minneapolis-St.Paul, Kansas City, Portland, Indianapolis, and Houston; 9th decil: New Orleans, and Cincinnati; 8th decil: Milwaukee; 7th decil: Buffalo, and Providence.
  Cities with 250,000–500,000 inhabitants: 10th decil: Hartford, Omaha, Richmond, Oklahoma City, Miami, Denver, Columbus, and San Diego; 9th decil: Nashville, San Antonio, Birmingham, Syracuse, Fort Worth, Louisville, and Memphis; 8th decil: Albany-Troy, and Springfield-Holyoke; 6th decil: Norfolk-Portsmouth, Rochester, and Toledo; 4th decil: Akron, and Dayton; 3rd decil: Youngstown; 2nd decil: Wilkes-Barre.

TABLE 14. Correlation Table: Finance, Insurance and Real Estate/Other Retail Trade. I–X, FIR-decils. A–J, Other Retail Trade, decils. For calculation of decils, see page 14 and Figures 29 and 30.

| | I | II | III | IV | V | VI | VII | VIII | IX | X | No. of cities |
|---|---|---|---|---|---|---|---|---|---|---|---|
| A | 56 | 15 | 6 | 3 | 3 | 1 | 0 | 2 | 0 | 0 | 86 |
| B | 15 | 20 | 14 | 12 | 4 | 4 | 4 | 8 | 2 | 4 | 87 |
| C | 5 | 16 | 17 | 11 | 6 | 10 | 2 | 5 | 9 | 5 | 86 |
| D | 4 | 9 | 16 | 14 | 14 | 7 | 7 | 5 | 6 | 5 | 87 |
| E | 5 | 8 | 12 | 16 | 8 | 12 | 5 | 9 | 3 | 8 | 86 |
| F | 0 | 10 | 12 | 12 | 12 | 11 | 9 | 4 | 7 | 10 | 87 |
| G | 0 | 4 | 3 | 8 | 18 | 10 | 10 | 7 | 12 | 14 | 86 |
| H | 0 | 4 | 5 | 7 | 9 | 14 | 17 | 11 | 7 | 13 | 87 |
| I | 0 | 1 | 1 | 1 | 9 | 11 | 14 | 17 | 20 | 12 | 86 |
| J | 1 | 0 | 0 | 3 | 3 | 7 | 19 | 18 | 20 | 15 | 86 |
| No. of cities | 86 | 87 | 86 | 87 | 86 | 87 | 87 | 86 | 86 | 86 | 864 |

twice as many people employed in retail trade as a typical city of the same size in the Northeast zone, which belongs to the first decil. Even the casual observer from the Northeast will note the unproportionately large business center of typical Prairie-zone towns.

Other Retail Trade is a major industry in all American cities; it is a major city forming activity in the majority of towns west of the Mississippi River. In the two western zones with one-fourth of the American cities are located almost half of the 320 B- and C-type retail towns, and a large

TABLE 15. Correlation Table: Finance, Insurance, and Real Estate/Wholesale Trade. I–X, FIR-decils. A–J, Wholesale Trade, decils. For calculation of decils, see page 14 and Figures 25 and 30.

|   | I | II | III | IV | V | VI | VII | VIII | IX | X | No. of cities |
|---|---|----|-----|----|---|----|-----|------|----|---|---------------|
| A | 49 | 13 | 10 | 8 | 2 | 3 | 0 | 1 | 0 | 0 | 86 |
| B | 20 | 23 | 15 | 12 | 5 | 3 | 3 | 4 | 0 | 1 | 86 |
| C | 5 | 19 | 15 | 16 | 7 | 8 | 6 | 5 | 2 | 3 | 86 |
| D | 4 | 4 | 15 | 13 | 11 | 13 | 8 | 6 | 7 | 5 | 86 |
| E | 2 | 9 | 11 | 13 | 14 | 8 | 8 | 8 | 6 | 8 | 87 |
| F | 3 | 9 | 6 | 8 | 18 | 12 | 5 | 10 | 6 | 9 | 86 |
| G | 1 | 6 | 4 | 6 | 6 | 13 | 16 | 15 | 11 | 9 | 87 |
| H | 1 | 1 | 6 | 3 | 10 | 13 | 13 | 10 | 20 | 10 | 87 |
| I | 1 | 0 | 3 | 4 | 9 | 5 | 17 | 10 | 14 | 24 | 87 |
| J | 0 | 3 | 1 | 4 | 4 | 9 | 11 | 17 | 20 | 17 | 86 |
| No. of cities | 86 | 87 | 86 | 87 | 86 | 87 | 87 | 86 | 86 | 86 | 864 |

number of the remaining retail towns are found in the adjacent areas of the North-central and Southern zones (Fig. 29).

Retail towns are found in areas with a relatively high density of rural population, exemplified by the vegetable and fruit growing areas of California, the Lower Rio Grande Valley and Florida, and/or a relatively low density of cities, exemplified by the prairie states, North and South Dakota and Nebraska. The rural population in the trade area of retail towns is not always predominantly agricultural, as indicated by several cities in the Appalachian coal field.

## Other Service Industries

### FINANCE, INSURANCE AND REAL ESTATE

This industry, here abbreviated to FIR, is with Business Service the most typical big-city activity. No other industries have such a large share of their employees living in large cities (Fig. 1). All cities with one-half million inhabitants or more have a considerably above-average value for FIR-employment (Table 13). Seven of twelve cities with more than one million inhabitants and eight

of thirteen with one-half to one million inhabitants belong to the tenth decil. A strong concentration in the highest decils is shown in Table 13 for cities with more than 100,000 inhabitants, an even distribution on the decils for cities with 25 to 100 thousand inhabitants, and a certain under-representation in the highest decils for the small cities with 10 to 25 thousand inhabitants.

There is a correlation of typological significance between the FIR-services on the one hand and retail trade (Table 14) and wholesale trade (Table 15), respectively, on the other. The latter two industries are strongly correlated between themselves (page 96). A typical American trade center thus belongs to the highest decils for all three of these industries. As it is more common for cities in the areas west of the Mississippi and in the South to function chiefly as trade centers than it is for cities in the Manufacturing Belt to do so, an unproportionately large number of towns with a high FIR-rate are found in the first two areas.

Since the FIR-services constitute a rather small industry which is highly ubiquitous in its distribution, they rank as a leading cityforming industry only in relatively few cities. There are just one city of B-type and fourteen of C-type.

Hartford, Connecticut, the most pronounced FIR-city in the United States, is the leading American life and fire insurance center, one of the largest insurance centers in the world.[1] It is unusual that a relatively small city and not the economic capital is the insurance center of a country. The insurance business of Hartford began in 1794.

FIG. 30. Cumulative distribution diagram for finance, insurance and real estate.

Of the fourteen FIR-cities of B-type, New York is outstanding. For a long time it has been the predominant financial center of the United States and after World War I it overtook London as the leading financial center of the world. Aside from apparel manufacturing the FIR-services are the leading cityforming industry in New York and the chief indicator of New York's position as the economic capital of the United States. The large concentration of financial activities in New York is the result of the paramount status of the city in American commerce and shipping, which was already reached in the early half of the nineteenth century with the opening of the Erie Canal (see page 99). A tremendous amount of economic power is concentrated in the city. Not only are the headquarters of leading banks located in New York but also those of huge companies in the oil, steel and movie industries, to mention but a few, although the actual production of these firms may

1. White and Foscue state that Hartford is the most important insurance center in the world (*Regional Geography of Anglo-America*, New York, 1943, p. 409).
2. The others are in Boston, New York, Philadelphia, Cleveland, Richmond, Atlanta, Chicago, St. Louis, Minneapolis, Kansas City, and San Francisco.

take place in other parts of the United States and abroad. The New York Stock Exchange, founded in 1792, plays a leading role in the American and international capital markets, just as the Cotton Exchange, the Coffee Exchange, the Cacao Exchange and others do in the commodity markets. Wall Street on lower Manhattan, around which much of the financial activity is concentrated, is one of the best known streets in the world.

Three of the other FIR-cities are primarily insurance centers: Des Moines, Iowa; Bloomington, Illinois; and Stevens Point, Wisconsin. Omaha, like the other large trade centers west of the Mississippi, is an important, diversified financial center. In addition to the usual activities, large insurance companies have their headquarters in the city. Dallas is the leading financial center of Texas and the Southwest. One of the twelve Federal Reserve Banks is located in the city.[2] Diversified FIR-functions are probably characteristic of Boise, Idaho; Helena, Montana; Fort Scott, Kansas; Owatonna, Minnesota; Wausau, Wisconsin; and Monmouth, Illinois. The location of Newport Beach, California, and Hollywood, Florida, in two outstanding recreation areas indicates a great importance for the real estate business in these cities.

BUSINESS SERVICE

Business service, the smallest urban industry specified by the American census, is the most urban activity (90.8 per cent urban) and the most pronounced big-city activity in the United States, with 76.8 per cent of its gainfully employed population living in cities of 100,000 or more inhabitants (Fig. 1). Of the total gainfully employed population in the United States 48.2 per cent live in such cities. Advertising, accounting, auditing, bookkeeping and other business services are associated with the headquarters of large firms and organizations, located in the big cities, and not with the branch offices in smaller towns. There are no business service towns in the United States.

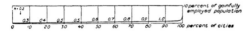

FIG. 31. Cumulative distribution diagram for business service.

## REPAIR SERVICES

Repair services constitute the most rural of American service industries. Only 66.3 of the employees are urban, as compared to 68.3 for the total gainfully employed population. Figure 1 indicates that most of the rural[1] population of this industry live in towns of less than 2,500 inhabitants. Also towns with 2,500 to 25,000 inhabitants have a larger percentage of repair service employees than would correspond to their share of the total employment. The rural orientation of the repair services is a result of the mechanization of farming and the importance of highway traffic. The average American farm with its large outfit of machines is like a small factory. Crossroad towns in farm areas have an unproportionately large number of repairmen. There are no repair service cities in the United States.

FIG. 32. Cumulative distribution diagram for repair service.

## PRIVATE HOUSEHOLDS

Even if the family is not primarily an association with economic goals, it should be reasonable to see the private household as a production unit, "the largest workshop in the country," producing food, lodging, etc.,

for the members of the family. But housewives, the largest category in household work, do not qualify as "gainfully employed." According to the concepts of the economists their housework does not contribute to the national income. From the geographical point of view the ubiquitous housewives are of little interest for an interpretation of the population distribution. The incompleteness of the private household "industry" in the census reports, accounting only for the relatively few employees but not for the many employers, is thus of little consequence for this study.

It is obvious that the private household "industry" is cityserving according to our terminology. The great variation in the private household percentages among American cities[2] is primarily a reflection of two ways of life in the United States, one typical for the South and the other for the rest of the country. In the South, with a large low-wage population of Negroes, it is still much more common to keep personal servants than in other parts of the country. A typical Southern town has more than six times as many household servants as an average New England town of the same size. Our assumption of an American standard in the consumption of services, of importance for the comparability of the relative maps, is not realistic for the private household "industry."

Some minor differences in the percentages for private household employment can be attributed to the different functions of cities. Professional towns, resorts, etc., have a somewhat higher percentage than manufacturing towns. Newport, Rhode Island, a wealthy summer resort, Asbury Park, New Jersey, also a summer resort, Princeton,

1. Rural according to the Census means towns smaller than 2,500 inhabitants and pure countryside.
2. Ten per cent of the cities have 1.5 per cent or less of their gainfully employed population engaged in private households, ten per cent have 8.6 per cent or more employed in this way (Fig. 33).

Private Household Service "Towns"

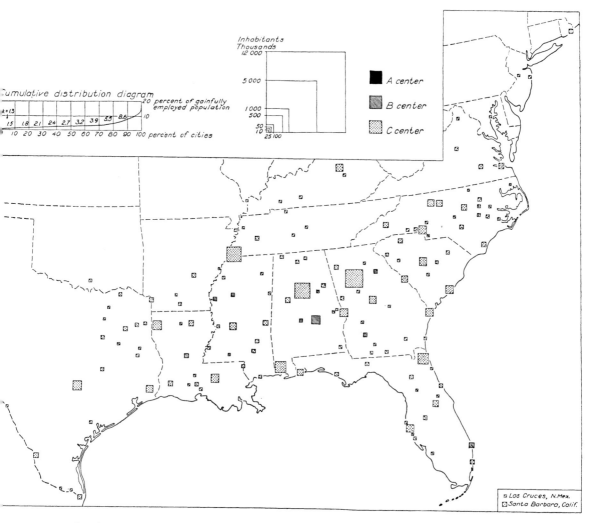

FIG. 33. Southern cities usually have 5–10 times as many people employed in private households as cities of the same size in New England.

New Jersey, a university town, Santa Barbara, California, a residential and resort city, and Las Cruces, New Mexico, are the only towns of C-type of this "industry" outside of the South.

HOTELS AND LODGING PLACES

Hotels and lodging places, one of the smallest American industries, is an important

cityforming activity only in a few towns: three of B-type and five of C-type.

Atlantic City, New Jersey, is the largest of the hotel towns. It is one of the world's largest summer resorts and a leading convention city. Its more than 1,000 hotels can accommodate over half a million guests. Atlantic City started to grow as a resort with the coming of the railroad about a hundred years ago. The sandy coast between Cape

Hatteras and the Hudson River has many smaller resorts. The dominance of Atlantic City is largely a result of its nearness to New York and Philadelphia and other big cities in northeastern United States, from which it is reached by excellent roads and fast railways. Points further south on the coast have a less favorable location for the east-west traffic because of the Delaware and Chesapeake bays.

FIG. 34. Cumulative distribution diagram for hotels and lodging places.

The two other hotel towns of B-type are Hollywood, Florida, a small winter resort north of Miami, and the gambling and divorce town, Las Vegas, Nevada.

Three of the hotel towns of C-type are resorts: Asbury Park, New Jersey, St. Augustine, Florida, the oldest city in the United States, founded by the Spanish in the sixteenth century, and Hot Springs, Arkansas, a well-known health and pleasure resort with a large number of mineral hot springs. The remaining two are college towns: State College, Pennsylvania, and Corvallis, Oregon.

OTHER PERSONAL SERVICES

Other personal services, comprising laundering, cleaning and dyeing services, dressmaking shops, shoe repair shops, etc., is one of the most ubiquitous American industries (Fig. 35). The health and recreation resort Hot Springs, Arkansas, is the only city with a remarkably high percentage for this industry (C-type).

FIG. 35. Cumulative distribution diagram for other personal services.

ENTERTAINMENT

The entertainment industry is a highly ubiquitous activity, engaging about one per cent of the gainfully employed population in most American cities (Fig. 36). There are only one entertainment city of B-type and two of C-type.

Las Vegas, Nevada, the most pronounced entertainment city in the United States, has a unique industrial structure. It has ratios of B-type for entertainment and hotels and of C-type for restaurants. Like Reno, the other "large" city in Nevada, an entertainment city of C-type, Las Vegas is known to newspaper readers all over the world for its gambling casinos and divorce courts. The rapid growth of the two Nevada cities in the last decades is chiefly the consequence of unique state laws which permit gambling of all types (since 1931) and easy divorce.

FIG. 36. Cumulative distribution diagram for entertainment.

The third entertainment city, Sarasota, Florida, is a winter resort. It is the winter headquarters of the world-famous Barnum and Bailey Circus.

Los Angeles has the highest entertainment percentage among the American metropolises (Appendix 2) thanks to the concentration of moving-picture production in the city. This branch of the entertainment industry is more concentrated geographically than any manufacturing branch. Probably nine-tenths of the American motion picture output has come from Los Angeles.[1] The industry is especially associated with Holly-

1. Clifford M. Zierer, "Hollywood—World Center of Motion Picture Production," *Annals of the American Academy of Political and Social Sciences* (November, 1947).

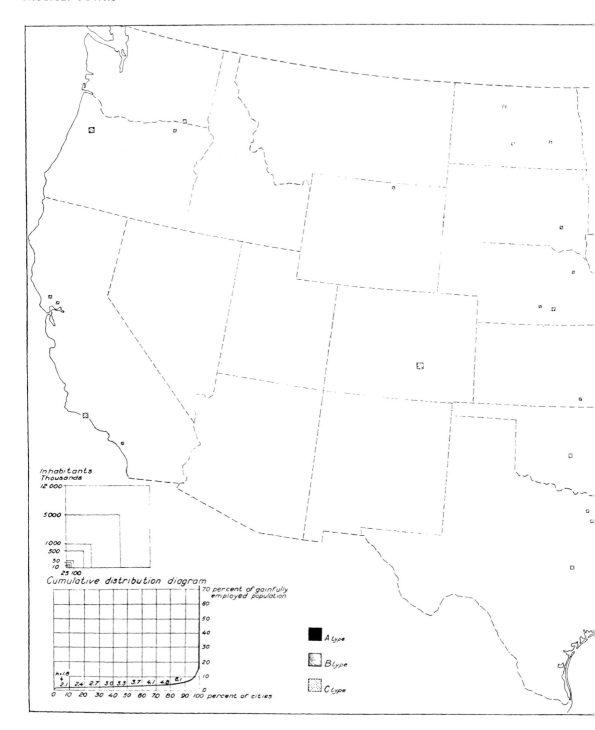

FIG. 37. Large hospitals usually are located in large cities. As a rule the only big hospitals located in smaller cities are of a special type, state hospitals for the insane, etc. There is one exceptional, specialized hospital town of a general type: Rochester, Minnesota. Thus American hospital towns usually are rather small, of B- or C-type, and scattered over the continent.

wood, a suburb of Los Angeles and one of the most widely known geographic names in the United States.

During its early years the moving-picture industry was located in the New York region, where the basic inventions had been made. Several firms in the beginning worked on roof-tops, where the light conditions were satisfactory. Edison's first studio for peep-show movies was located in West Orange, a suburb of New York. When the film story became popular around 1903, producers in New York and Chicago, the second producing center, sent expeditions to the West to get authentic backgrounds for "Wild West" movies, which were popular from the beginning. Other expeditions were sent to Florida and to Cuba to get sunshine and scenery. For several years Jacksonville, Florida, had a large movie colony during the winter months.[1]

The uniformity and mildness of the climate of Southern California, as well as the great variety of scenery, early attracted motion picture producers from New York and Chicago. The first studio in the Los Angeles region was built in 1909 and the first one in Hollywood in 1911. The subsequent movie boom is reflected in the population growth of Hollywood, incorporated in Los Angeles in 1910 with 5,000 inhabitants. Ten years later the suburb had 36,000 inhabitants and in 1930 not less than 235,000. Several of the largest studios, however, were built outside of Hollywood in other parts of the Los Angeles area, especially in the San Fernando Valley.

With the coming of sound at the end of the twenties the industry made increased use of indoor "shooting," which lessened the importance of climate and scenery as locational factors. Los Angeles, however,

was by then deeply entrenched as the production center of the industry. Huge investments in plant and equipment and the geographic concentration of movie specialists, actors, camera men, writers and others became the chief advantages of the city. But it can be safely stated that the movie industry would not have left the Manufacturing Belt if it had not gone through its "outdoor" and "western" stages of development.

New York has remained the seat of distribution and financing,[2] the headquarters for the large movie companies.

MEDICAL SERVICES

The medical cities are scattered over the United States roughly in proportion to the distribution of cities, but with some under-representation in the Manufacturing Belt with its high city density (Fig. 37). Medical cities in the United States are small, with only one, Lexington, Kentucky, reaching the size of 100,000 inhabitants. In 70 percent of the cities the medical service percentage varies within the narrow range of 2.1 to 4.8. General hospitals and general practitioners have to be easily accessible for the inhabitants of a city and its surrounding countryside. Such services will therefore as a rule be located in the city (Fig. 1), the hub of transportation of its trade area. People living in small towns or in the countryside will usually go to the doctor or to the hospital in the city which they refer to as "town." The service area of the hospitals will roughly coincide with the trade area of the city.

Cities with a large enough medical service percentage to make them medical cities generally have one or more special types of hospitals, like state hospitals for the insane, veterans hospitals, general hospitals connected with a college of medicine in small

1. *Ibid.*
2. J. G. Glover and W. B. Cornell, *The Development of American Industries* (New York, 1946), p.811.

8 — 565447 *Alexandersson*

university towns (type: Iowa City), etc.
There are only three medical cities of A-
type in the United States and eleven of B-
type.

The metropolises with more than one
million inhabitants usually have a medical
service percentage close to the median
average for cities with more than 10,000
inhabitants. Boston has the highest rate of
these cities (Appendix 2).

Rochester, Minnesota, with the world-
famous Mayo Clinic, is a unique medical
city. The Mayo medical center was organized in 1889 by Doctor W. W. Mayo and
his sons. It was the outgrowth of a hospital
that had been founded a few years earlier
after a tornado had swept the town. In 1915
the Mayo Foundation for Medical Education and Research was created. Affiliated
with University of Minnesota, it draws
hundreds of students to Rochester from all
over the country for graduate work in
medicine. In 1924 the Mayo Institute of
Experimental Medicine was opened. Rochester has 1600 hospital beds in six hospitals.
The State Hospital for the Insane, located
just east of the city, has an average of 1600
patients.[1]

Both Kings Park, New York, and Midway-Hardwick, Georgia, the other two
medical towns of A-type, are unincorporated
places which reach the size of 10,000 inhabitants only because the patients of their
big hospitals are included. They are extremely one-sided, with 74 and 66 per cent,
respectively, of their gainfully employed
population engaged in medical services.

EDUCATION

Education is a major industry everywhere
in the United States (Fig. 1 and Fig. 38).
Grade schools are ubiquitous in both rural
and urban areas. High schools are available
even in small towns and are within school

bus range of most American homes. The
number of teachers and other personnel in
primary and secondary schools is roughly
proportionate to the population of the urban
places. It is thus the institutions of higher
education, of necessity larger and more
specialized, which constitute the city forming element in the educational industry.
Higher education in the United States differs from that of European countries in two
conspicuous ways. The enrollment, and
therefore the teaching staff, is proportionally
much larger. The institutions are often
located in small cities.

Contrary to the European countries,
where college education is the privilege of
a small, selected group of students, predominantly recruited from upper and middle
class families, the American educational
system ambitiously attempts mass education
even at the university level.

In the United States many colleges ·and
universities were deliberately located in
small towns, whereas in Europe leading
cities were chosen as locations for universities. Such relatively small university cities
as Coimbra, Portugal, Uppsala and Lund,
Sweden, and Cambridge and Oxford, England, were old capitals, bishop seats, etc.,
with greater relative importance at the time
when the university was founded than at
present.

Even the big-city universities in the
United States were often given a rural
orientation. They are as a rule located in
the outskirts of the present urbanized area,
often several miles from the downtown district. Here land was available for the large
campus with its many huge buildings for
teaching and research departments, dormi-
tories, a stadium, etc. A large American
university has ten to fifteen thousand students enrolled.

1. M. I. Fletcher, "Rochester: A Professional
Town," *Economic Geography* (1947).

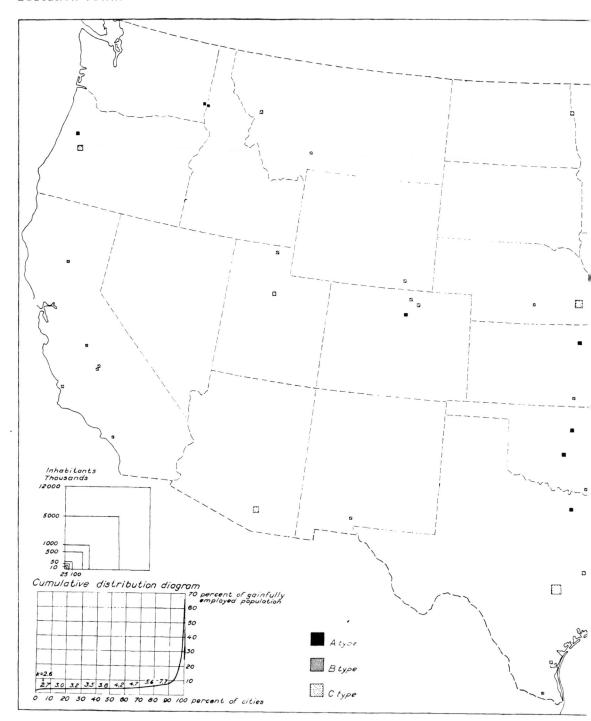

FIG. 38. Many large American universities are located in small cities. Thus education towns constitute an important functional type of cities in the United States.

The rural orientation of many institutions of higher education in the United States should be seen as the result of ideas which prevailed during a relatively short period at the end of last century when most colleges and universities were founded or substantially enlarged. Life on the farm or in the small town was preferred to the "wicked city" by the university founders, many of whom were first-generation city dwellers or country people. Because of the handicaps to professional education, notably medicine, engineering, and business, and because of limited employment for impecunious students, it is unlikely that new institutions will be located in small towns.[1]

Most early American schools were private rather than public, partly because of the prevalent laissez-faire philosophy, partly because of the financial inability of the pioneer society to provide democratic education for everybody. Some of the best-known American universities are private institutions: Harvard, Yale, Columbia, Princeton, John Hopkins, the University of Chicago. Large gifts to institutions of higher education from private philanthropists became increasingly common during the period of rapid business expansion after the Civil War. The total endowment of colleges and universities in the United States reaches enormous figures.

The government also became actively engaged in supporting higher education at this time. The Morrill Act of 1862 marks a new era in this respect. Under its terms the federal government gave generous amounts of land to the states if they established at least one college. Ultimately, not less than sixty-nine "land-grant colleges" profited

from the terms of the Act.[2] Some of these, like the state universities of California, Wisconsin, Illinois, and Minnesota, are among the largest and most influential universities in the country.

Among the leading institutions of higher education in the different states are:

*Maine:* University of Maine at Orono[3]; *New Hampshire:* Dartmouth College at Hanover* and University of New Hampshire at Durham*; *Vermont:* University of Vermont and the State Agricultural College at Burlington; *Massachusetts:* Harvard University and Massachusetts Institute of Technology at Cambridge (Boston); *Rhode Island:* Brown University at Providence; *Connecticut:* Yale University at New Haven; *New York:* Columbia University and New York University in New York, Cornell University at Ithaca, Syracuse University; *New Jersey:* Princeton University and the Institute for Advanced Study at Princeton, Rutgers University at New Brunswick (New York); *Pennsylvania:* University of Pennsylvania in Philadelphia, the University of Pittsburgh, Pennsylvania State College at State College; *Maryland:* John Hopkins University in Baltimore, University of Maryland at College Park (Washington), U.S. Naval Academy and St. Johns College at Annapolis; *West Virginia:* West Virginia University at Morgantown; *Ohio:* Ohio State University at Columbus; *Michigan:* University of Michigan at Ann Arbor, Wayne University in Detroit, Michigan State College of Agriculture and Applied Science at Lansing; *Indiana:* Indiana University at Bloomington, Purdue University at Lafayette; *Illinois:* University of Chicago, Northwestern University at Evanston (Chicago), University of Illinois at Champaign-Urbana; *Wisconsin:* University of Wisconsin at Madison; *Minnesota:* University of Minnesota in Minneapolis; *Iowa:* State University of Iowa at Iowa City, Iowa State College of Agriculture and Mechanic Arts at Ames; *Missouri:* University of Missouri at Columbia, St. Louis University; *North Dakota:* University of North Dakota at Grand Forks; *South Dakota:* University of South Dakota at Vermillion*; *Nebraska:* University of Nebraska at Lincoln; *Kansas:* University of Kansas at Lawrence, Kansas State College of Agriculture and Applied Science at Manhattan; *Virginia:* College of William and Mary at Williamsburg*, University of Virginia at Charlottesville; *North Carolina:* University of

1. S. A. Queen and L. F. Thomas, *The City* (New York, 1939), p. 138.

2. John D. Hicks, *The American Nation* (Cambridge, 1949), p. 99.

3. An asterisk marks towns of less than 10,000 inhabitants.

North Carolina at Chapel Hill (Durham), State College of Agriculture and Engineering at Raleigh; *South Carolina:* Allen University of South Carolina at Columbia; *Georgia:* University of Georgia at Athens; *Florida:* University of Florida at Gainesville, Florida State University and Florida Agricultural and Mechanical College at Tallahassee; *Alabama:* University of Alabama at University (Tuscaloosa), Alabama Polytechnic Institute at Auburn; *Mississippi:* University of Mississippi at Oxford*; *Louisiana:* Louisiana State University at Baton Rouge; *Kentucky:* University of Kentucky at Lexington; *Tennessee:* University of Tennessee at Knoxville; *Arkansas:* University of Arkansas at Fayetteville; *Oklahoma:* University of Oklahoma at Norman, Oklahoma Agricultural and Mechanical College at Stillwater; *Texas:* University of Texas at Austin, Baylor University at Waco, Southern Methodist University at Dallas, North Texas State College and Texas State College for Women at Denton; *New Mexico:* University of New Mexico at Albuquerque; *Arizona:* University of Arizona at Tucson; *Nevada:* University of Nevada at Reno; *Utah:* University of Utah at Salt Lake City, Brigham Young University at Provo; *Colorado:* University of Colorado at Boulder; *Wyoming:* University of Wyoming at Laramie; *Montana:* Montana State University at Missoula, Montana State College at Bozeman; *Idaho:* University of Idaho at Moscow; *Washington:* University of Washington at Seattle, State College of Washington at Pullman; *Oregon:* Oregon State College at Corvallis, University of Oregon at Eugene; *California:* University of California at Berkeley (San Francisco) and in Los Angeles, University of Southern California in Los Angeles, Stanford University at Palo Alto (San Francisco).

### OTHER PROFESSIONAL AND RELATED SERVICES

Other professional and related services, comprising welfare, religious and membership organizations, legal, engineering and architectural services, etc., is another highly ubiquitous industry (Fig. 39). The university

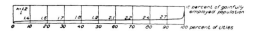

FIG. 39. Cumulative distribution diagram for other professional services.

town Princeton, New Jersey, is the only city with a C-type dominance in this industry.

### PUBLIC ADMINISTRATION

Public administration comprises all forms of federal, state and local government. In this group are included employees of the postal service and the armed forces. Public administration is a leading cityforming activity chiefly in cities of two types: political capitals and cities with military installations. There is no conspicuous regional concentration of cities with high or low percentages for public administration with the exception of the West zone, where nine-tenths of the cities have an above average rate and one-half belong to the two highest decils. The chief reason for the unusually high percentage of government employees in most Western cities seems to be the large irrigation projects which are typical of this area and which usually are in charge of the federal Bureau of Reclamation. Cities with more than 250,000 inhabitants have an above-average public administration percentage with only few exceptions;[1] for smaller cities no clear tendency is discernable.

1. *Cities with more than 1,000,000 inhabitants:* *10th decil:* Washington and San Francisco; *9th decil:* Boston and Baltimore; *8th decil:* St. Louis and Los Angeles; *7th decil:* Philadelphia and New York; *6th decil:* Cleveland, Chicago and Pittsburgh; *4th decil:* Detroit.
  *Cities with 500,000–1,000,000 inhabitants:* *9th decil:* Seattle, Atlanta and New Orleans; *8th decil:* Portland, Kansas City, Minneapolis-St. Paul, Indianapolis and Providence; *7th decil:* Dallas; *6th decil:* Buffalo, Milwaukee and Cincinnati; *2nd decil:* Houston.
  *Cities with 250,000–500,000 inhabitants:* *10th decil:* Norfolk-Portsmouth, San Antonio, San Diego, Oklahoma City, Albany-Troy, Dayton and Columbus; *9th decil:* Denver and Richmond; *8th decil:* Hartford and Nashville; *7th decil:* Omaha, Louisville, Springfield, Port Worth, Miami and Toledo; *6th decil:* Memphis; *5th decil:* Rochester; *4th decil:* Wilkes-Barre; *3rd decil:* Birmingham; *2nd decil:* Youngstown and Akron.

## Political Capitals

Washington, the federal capital, is the dominant political and administrative center of the United States. Not less than 32 per cent of its gainfully employed population are government employees.

The city is to be seen as a product of rivalry between cities and states in the early years of the nation. When the Northern and Southern states could not agree on any important city for the national capital, a compromise was reached according to which Virginia and Maryland set aside land on both sides of the Potomac for a federal district. The Virginia part was later returned to the state and the present District of Columbia is located on the Maryland side of the river. The urbanized area of Washington now includes large suburbs in Virginia and Maryland. In 1950 it had 1.3 million inhabitants with 800,000 in the District of Columbia.

The new capital had a central position at the time with respect to the population distribution. The site on the Fall Line, at the head of navigation on the Potomac River, was considered to be favorable, since the river was expected to become the big highway to the Middle West.[1] The dreams of a future for Washington as a center of commerce and manufacturing did not come true, however, but the ever increasing activities of the federal government have proved to be a sufficient basis for a large metropolis. The huge government buildings, mainly neoclassic in style, are given a dominant position in the center of the city, which has been developed according to L'Enfant's grandiose city plan. They give a distinct character to Washington, making it different from other big cities in the United States and the world.

1. Ralph H. Brown, *Historical Geography of the United States* (New York, 1948), p. 140.

Public administration is by far the leading industry in Washington (Appendix 2). As a consequence of the concentration of political and administrative activities in the city, a great variety of organizations have found it expedient to locate their headquarters here or at least to maintain an office in Washington. Such interest groups, lobbies, are common in all political capitals of democratic countries. In Washington "lobbying" is an important cityforming activity although it is not represented under any single industry group in the statistics.

For a city of its size Washington is extremely unimportant in manufacturing. It has the lowest percentages of the twelve cities with more than one million inhabitants for all manufacturing industries except printing and publishing—in which it has the highest—and construction. As a commercial center Washington is relatively unimportant with the lowest wholesale trade percentage of forty-nine cities with more than 250,000 inhabitants.

The state capitals vary considerably in size. Four have less than 10,000 inhabitants: Carson City, Nevada, Pierre, South Dakota, Montpelier, Vermont, and Dover, Delaware. Others, like Boston and Atlanta, are the biggest cities of their states. In all capitals public administration is a major cityforming activity, but only one with more than 10,000 inhabitants, Olympia, Washington, is a one-sided administration city. Fig. 39 shows all state capitals. Their names are in capital letters.

## Military Towns

Like the political capitals, the military towns can derive their relatively dominant activity from a precise origin, a legislative act. Their military function has not developed gradually as their manufacturing, trade or transportation functions. Even if local interests in the past may have influenced the location

of Army posts and Naval bases by political
manipulation in Congress,[1] military installations are as a rule located with regard to the
needs of national defense and are planned
and built by a central authority. Of the
three branches of the armed services, the
Navy and the Air Force are characterized
by very large installations, bringing about a
concentration in fewer cities and a strong
cityforming influence even in relatively
large urban centers.

Whereas the choice of sites for naval
bases is limited by the relatively small
number of suitable ports, Army posts and
Air Force bases can theoretically be located
anywhere within very large areas. Locations
not too far from established cities seem,
however, to be preferred: there is no
example in the United States of cities having
grown to the size of 10,000 or more inhabitants exclusively as military towns—
like Wilhelmshaven, Germany, or Karls-
krona, Sweden. For various reasons—
favorable flying weather, cheap land for
extensive testing and training grounds, etc.
—the states west of the Mississippi have a
disproportionately large share of the American Air Force and Army garrisons. The vast
desert and steppe areas of the western
United States have been the scene of many
tests with atomic weapons, and large guided-
missile experiment stations are located here
(for example, Albuquerque and El Paso).

The largest naval installations on the
Atlantic coast and the headquarters of the
Atlantic fleet are in the Hampton Roads

cities, Norfolk-Portsmouth and Newport
News. The wealthy summer resort, Newport, Rhode Island, is an important Navy
base. Other cities on the East Coast with
large naval installations are Jacksonville and
Key West, Florida. Huge naval air stations
have made great impacts on the industrial
structures of Pensacola, Florida, and Corpus
Christi, Texas.

On the Pacific coast, San Diego, with one
of the few natural ports in California, is the
leading naval base. San Francisco, an
administrative city of C-type, has naval
installations in addition to a large number of
state and federal activities. Oxnard in
southern California is a one-sided military
city.

Several large military cities of B- and C-
type are chiefly Air Force bases: San
Antonio, Texas, Dayton, Ohio (home town
of the Wright brothers); San Bernardino,
California; Albuquerque, New Mexico;
Macon, Georgia; and Mobile, Alabama.
Tacoma, Washington, and El Paso, Texas,
have Army forts as well as Air Force bases.
Ogden, Utah, an administration city of A-
type and a railroad center of B-type, has a
U.S. ordnance depot, railroad shops of the
Transportation Corps and an air base.
Ogden, Key West, Florida, and Oxnard,
California, were the only military cities of
A-type in the United States with more than
10,000 inhabitants in 1950.

1. C. B. Swisher, *The Theory and Practice of
American National Government* (New York, 1951),
p. 795.

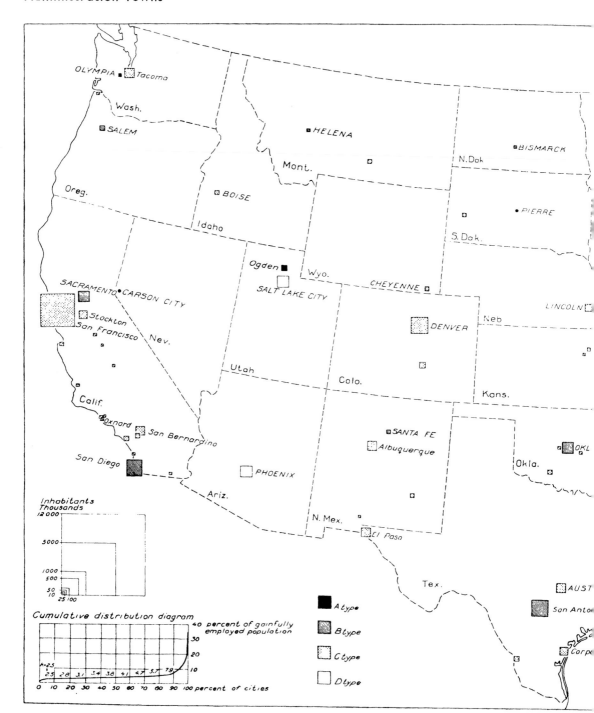

FIG. 40. The names of state capitals are capitalized on the map. State capitals with less than 10,000 inhabitants are shown with small circles. Administration towns other than political capitals have as a rule military functions.

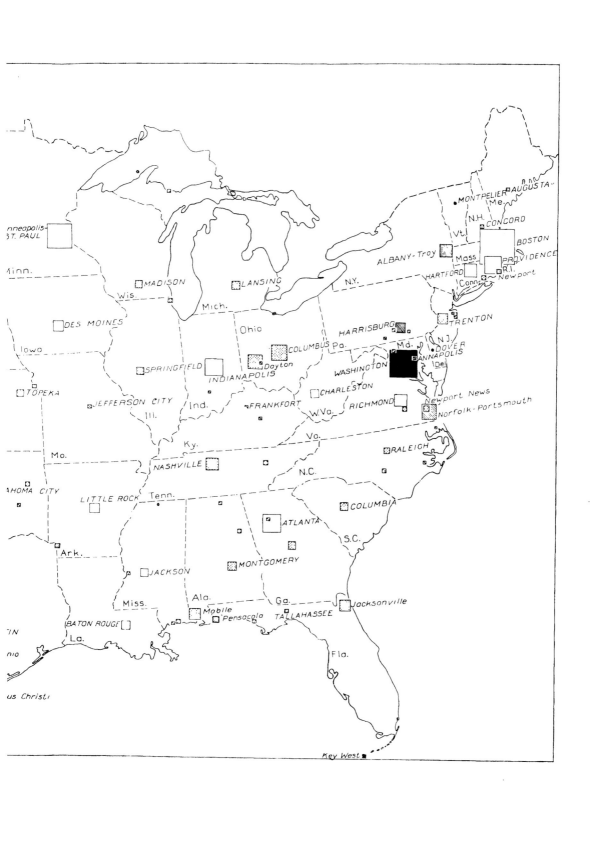

# APPENDIX 1. Chief City Forming Industries in American Cities with 10,000 or More Inhabitants.

Inhabitants in thousands in parenthesis after the name of the city.

A-type industries employ $k + 20.0$ per cent or more of the gainfully employed population of the city. The small rural population was not considered.

B-type industries employ $k + 10.0$ to $19.9$ per cent of the gainfully employed population.

C-type industries employ $k + 5.0$ to $9.9$ per cent of the gainfully employed population.

For the calculation of the constant $k$, see pages 17 ff., and Figures 2–40.

For a fuller definition of the industries, see Fig. 1. The word "Other" has been left out in this list for Other Nondurable Goods, Other Durable Goods, Other Retail Trade, etc.

### MAINE

| | |
|---|---|
| Bangor (32) | C. *Retail, Nondurable, Medical, Wholesale, Railroads* |
| Waterville (18) | B. Textile, Nondurable |
| | C. *Railroads* |
| Augusta (21) | B. Nondurable, Textile, *Administration* |
| | C. *Medical* |
| Lewiston-Auburn (64) | A. Textile, Nondurable |
| Bath (11) | A. Transport. equipment |
| Portland (113) | C. Nondurable, *Wholesale, Retail* |
| Biddeford-Saco (31) | A. Textile, Machinery |
| | C. Nondurable |
| Sanford (11) | A. Textile |
| | B. Nondurable |

### NEW HAMPSHIRE

| | |
|---|---|
| Berlin (17) | A. Nondurable |
| Laconia (15) | B. Machinery |
| | C. Textile |
| Claremont (13) | A. Machinery |
| | B. Nondurable |
| | C. Textile |
| Concord (28) | C. Printing, *Administrat., Medical, Railroads* |
| Rochester (14) | A. Nondurable |
| | B. Textile |
| | C. Furniture & Lumber |
| Dover (16) | B. Nondurable, El. mach. |
| | C. Textile |
| Portsmouth (19) | B. Transport. equipment |
| | C. Nondurable |
| Keene (16) | C. Textile, Furn. & Lumber, Nondurable, Machinery |
| Manchester (85) | A. Nondurable |
| | B. Textile |
| Nashua (35) | A. Nondurable |
| | B. Textile |

### VERMONT

| | |
|---|---|
| Burlington (33) | C. *Education*, Textile |
| Barre (11) | A. Durable |
| | C. El. machinery |
| Rutland (18) | C. *Railroads*, Machinery, *Retail* |

### MASSACHUSETTS

| | |
|---|---|
| Newburyport (14) | B. Nondurable, El. machinery |
| | C. Durable |
| Haverhill (47) | A. Nondurable |
| Lawrence (112) | A. Textile |
| | C. Nondurable |
| Lowell (107) | A. Textile |
| | C. Nondurable |
| Gloucester (25) | B. Food |
| | C. *Wholesale* |
| Boston (2 233) | C. Nondurable |
| Plymouth (11) | A. Textile |
| New Bedford (125) | A. Textile |
| | C. Apparel, El. machinery |
| Fall River (118) | A. Textile |
| | B. Apparel |
| Taunton (40) | B. Durable, Textile |
| | C. Nondurable |
| Brockton (92) | A. Nondurable |
| Webster (12) | A. Textile |
| | B. Nondurable |
| | C. Durable |
| Southbridge (17) | A. Durable |
| | B. Textile |
| | C. Fabricated metal |
| Milford (14) | A. Machinery |
| | B. Nondurable |
| | C. Textile |
| Marlborough (16) | A. Nondurable |
| Worcester (219) | C. Machinery, Primary metal, Durable, Nondurable |

Clinton (12) — B. Textile, Fabr. metal, Printing

Leominster (24) — A. Durable / C. Apparel

Fitchburg (43) — B. Nondurable, Fabricated metal / C. Durable, Textile

Gardner (20) — A. Furniture & Lumber / B. Durable / C. Fabricated metal

Springfield-Holyoke (367) — C. Nondurable, El. machinery, Machinery, Textile

Greenfield (15) — B. Machinery / C. *Railroads, Retail*, Fabricated metal

North Adams (22) — A. El. machinery, Textile

Adams-Renfrew (12) — A. Textile / B. El. machinery / C. Nondurable

Pittsfield (53) — A. El. machinery / C. Textile

RHODE ISLAND

Woonsocket (40) — A. Textile

Providence (583) — B. Textile, Durable

Bristol (10) — A. Nondurable / B. Textile / C. Primary metal

Newport (38) — B. *Administration* / C. *Households*

CONNECTICUT

Willimantic (14) — A. Textile / B. El. machinery / C. Fabricated metal

Norwich (23) — B. Textile / C. Fabricated metal, Nondurable

New London (31) — C. *Administration*, Transport. equipment

Hartford (301) — B. *Finance*, Machinery / C. Transport. equipment

Torrington (28) — B. Durable, Machinery, Fabricated metal / C. Primary metal

New Britain - Bristol (123) — B. Fabr. metal, Machinery / C. Durable, El. machinery

Waterbury (132) — B. Primary metal, Durable / C. Fabr. metal, Nondurable

Meriden (44) — A. Durable / B. Machinery

Middletown (30) — B. Textile / C. *Medical*, Machinery

Wallingford (12) — A. Durable / C. Fabricated metal

New Haven (245) — C. Fabr. metal, Apparel

Bridgeport (237) — B. Machinery / C. El. machinery, Fabr. metal, Primary metal

Ansonia-Derby (29) — B. Primary metal, Nondurable / C. Fabr. metal, Textile

Danbury (22) — A. Textile

Stamford-Norwalk (174) — C. Fabr. metal, Machinery

NEW JERSEY

Red Bank (13) — B. Apparel, *Administration*

Long Branch (23) — B. *Administrat.*, Apparel

Asbury Park (17) — C. *Administrat.*, Apparel, *Hotels, Households*

Atlantic City (105) — B. *Hotels* / C. Construction, *Eating*

Trenton (189) — C. Durable, Nondurable, *Administration*

Princeton (12) — A. *Education* / C. *Households, Prof. serv.*

Burlington (12) — B. Primary metal / C. Apparel, Durable

Millville (15) — A. Durable / B. Textile / C. Apparel

Bridgeton (18) — B. Durable / C. Apparel, Food

NEW YORK

New York (12 286) — C. Apparel, *Finance*

Kings Park (11) — A. *Medical*

Peekskill (18) — C. *Railroads*, Food, Apparel

Middletown (23) — B. *Medical* / C. Apparel, *Railroads*

Newburgh-Beacon (46) — B. Apparel, Textile

Beacon (46) — C. Nondurable, *Medical*

Poughkeepsie (41) — B. Machinery / C. Apparel

Kingston (29) — B. Apparel

Hudson (11) — B. Durable / C. Textile

Albany-Troy (291) — B. *Administration* / C. *Railroads*, Apparel

Schenectady (123) — A. El. machinery

Oneonta (14) — B. *Railroads* / C. *Retail*

Amsterdam (32) — A. Textile

Johnstown (11) — A. Nondurable / B. Textile

Gloversville (24) — A. Nondurable / C. Textile

Saratoga Springs (15) — C. Textile, *Education*

Glens Falls (20) — C. Nondurable, Apparel, *Retail*

Utica (117) — B. Textile

Rome (42) — A. Primary metal / C. *Medical*, El. machinery

Oneida (11) — A. Durable

Syracuse (265) — C. Machinery, El. machinery

Fulton (14) — A. Nondurable / B. Textile, Food

Oswego (23) — B. Nondurable / C. Fabricated metal

Auburn (37) — B. Machinery, Textile / C. Nondurable

Geneva (17) — C. Machinery, Fabricated metal, *Retail*

| | | | |
|---|---|---|---|
| Newark (10) | B. *Retail, Medical* | Sunbury (16) | B. Textile |
| | C. Nondurable | | C. El. machinery, *Rail-* |
| Cortland (18) | C. Prim. metal, Apparel, | | *roads*, Apparel |
| | Machinery | Easton, Pa. — | C. Machinery, Textile, |
| Ithaca (29) | A. *Education* | Phillipsburg, N.J. (55) | Apparel, Prim. metal |
| | C. Machinery | Allentown-Bethlehem | B. Primary metal |
| Binghamton (144) | A. Nondurable | (226) | C. Apparel, Textile |
| | B. Machinery | Bristol (13) | B. Chemicals |
| Elmira (50) | B. Machinery | | C. El. machin., Fabricat. |
| Corning (18) | A. Durable | | metal, Textile, Motor |
| | C. *Railroads* | | vehicles, Nondurable |
| Hornell (15) | A. *Railroads* | Philadelphia (2 922) | |
| | C. Textile | Conshohocken (11) | B. Prim. metal, Nondur- |
| Rochester (409) | B. Durable | | able |
| | C. Apparel | | C. Textile, Durable, |
| Batavia (18) | B. Machinery | | Fabric. metal |
| | C. Prim. metal, Nondur- | Norristown (38) | B. Textile |
| | able, *Medical* | | C. Prim. metal, Durable |
| Lockport (25) | A. Motor vehicles | Phoenixville (13) | B. Textile, Nondurable, |
| | C. Nondurable | | Primary metal |
| Niagara Falls (98) | B. Chemicals | | C. Fabr. metal, *Medical* |
| | C. Durable, Prim. metal, | Pottstown (23) | B. Primary metal |
| | Nondurable | | C. Fabr. metal, Apparel, |
| Buffalo (798) | C. Primary metal | | Motor vehicles, Non- |
| Dunkirk (18) | B. Prim. metal, Fabric. | | durable |
| | metal | Reading (155) | B. Textile |
| | C. Textile, Machinery | | C. Prim. metal, Machin. |
| Jamestown (43) | B. Furn. & Lumb., Fa- | West Chester (15) | B. *Education* |
| | bric. metal | | C. *Medical* |
| | C. Machinery, Textile | Coatesville (14) | A. Primary metal |
| Olean (23) | B. Machinery | | C. *Medical* |
| | C. Furniture & Lumber | Lancaster (76) | B. Textile |
| Watertown (34) | C. Machinery, *Retail*, | | C. Durable, El. machin. |
| | *Railroads* | Columbia (12) | B. Textile, Apparel |
| Ogdensburg (16) | B. *Medical* | | C. Prim. metal, *Admini-* |
| | C. Nondurable, Furn. & | | *stration*, Durable |
| | Lumber | York (79) | B. Machinery |
| Massena (13) | A. Primary metal | | C. Fabricated metal |
| Plattsburg (18) | B. Nondurable | Hanover (14) | B. Nondurable |
| | C. *Medical, Education,* | | C. Textile |
| | Fabricated metal | Lebanon (28) | B. Primary metal, Apparel |
| | | | C. Textile |
| | | Harrisburg (170) | B. *Administration* |
| PENNSYLVANIA | | | C. Primary metal, |
| | | | *Railroads* |
| Scranton (236) | B. Mining, Apparel | Carlisle (17) | B. Nondurable, |
| | C. *Railroads* | | *Administration* |
| Wilkes-Barre (272) | A. Mining | | C. Textile |
| | C. Apparel | Chambersburg (17) | B. *Administration* |
| Berwick (14) | A. Transport. equipment | | C. Apparel, Food |
| | B. Textile | Waynesboro (10) | A. Machinery |
| | C. Food, Apparel | State College (17) | A. *Education* |
| Bloomsburg (11) | A. Textile | | C. *Hotels* |
| | C. *Retail* | Lewistown (14) | A. Chemicals |
| Hazleton (35) | B. Mining, Apparel | | C. Primary metal, *Retail* |
| | C. Textile | Williamsport (45) | C. Nondurable, Textile |
| Tamaqua (12) | A. Mining | Lock Haven (11) | B. Nondurable, |
| | B. *Railroads*, Chemicals | | El. machin. |
| Mahanoy City (11) | A. Mining | Bradford (17) | B. Mining |
| | B. Apparel | | C. Nondurable, |
| Shenandoah (16) | A. Mining | | Machinery |
| | B. Apparel | Warren (15) | B. Fabricated metal |
| Pottsville (24) | C. Apparel, Mining, | | C. *Retail* |
| | *Retail* | Du Bois (11) | B. *Railroads*, Durable |
| Mount Carmel (14) | A. Mining | | C. *Retail* |
| | B. Apparel | Altoona (87) | A. *Railroads* |
| Shamokin (17) | B. Mining, Apparel, | Johnstown (93) | A. Primary metal |
| | Textile | | |

Indiana (12)    B. Mining
                C. *Education, Retail*
Jeannette (16)  A. Durable
                C. Machinery, Prim.
                metal
Latrobe (12)    A. Primary metal
Greensburg (17) C. Machinery, *Retail*,
                Durable
Connellsville (13) B. *Railroads*, Durable
                C. Mining
Uniontown (20)  B. Mining, *Retail*
Monessen (18)   A. Primary metal
Donora (12)     A. Primary metal
Washington (26) A. Durable
                C. Primary metal
Canonsburg (12) B. Primary metal
                C. El. machin., Fabric.
                metal, Mining, Dur-
                able
Pittsburgh (1 533) B. Primary metal
Ellwood City (13) A. Primary metal
                C. Machinery, Durable
New Castle (49) B. Durable, Machinery
                C. *Railroads*, Prim. metal,
                Fabricated metal
Butler (23)     B. Primary metal
                C. *Retail, Medical*
Sharon-Farrell (41) A. Prim. metal,
                El. machinery
Franklin (10)   A. Machinery
                C. Printing
Oil City (20)   B. Machinery
                C. Nondurable, *Railroads*
Meadville (19)  B. Durable, Chemicals
                C. *Railroads*
Erie (152)      B. Machinery, El. mach.
                C. Durable, Nondurable

DELAWARE

Wilmington (187) B. Chemicals
                C. Nondurable

MARYLAND

Cumberland (38) B. *Railroads*, Chemicals
                C. Nondurable
Hagerstown (36) B. Transport. equipment
                C. *Railroads*
Frederick (18)  C. *Administration*,
                Apparel, Construc-
                tion, *Retail*
Annapolis (10)  A. *Education*
                B. *Administration*
                C. *Households*
Baltimore (1 162) C. Primary metal
Cambridge (10)  B. Food
                C. Apparel, Construction
Salisbury (15)  C. *Retail*, Apparel, Food

DISTRICT OF COLUMBIA

Washington (1 287) A. *Administration*

WEST VIRGINIA

Weirton (24)    A. Primary metal
Wheeling (107)  B. Primary metal
                C. Mining, Durable

Moundsville (15) A. Durable
                C. Mining, Prim. metal,
                Fabricated metal
Morgantown (26) B. *Education*, Mining
                C. Durable
Fairmont (29)   B. Mining
                C. Durable
Clarksburg (32) B. Durable
                C. Mining, *Retail*
Parkersburg (40) B. Chemicals
                C. *Retail*, Durable
Martinsburg (16) B. Textile
                C. *Medical, Railroads*
Charleston (131) B. Chemicals
Huntington, W.Va.- B. *Railroads*
Ashland, Ky. (156) C. Primary metal
Beckley (19)    B. Mining
                C. *Retail*
Bluefield (22)  B. *Railroads*
                C. Mining, *Wholesale,
                Retail*

OHIO

Conneaut (10)   A. *Railroads*
                B. El. machinery
Ashtabula (24)  B. *Railroads*
                C. *Oth. transp.*,
                El. machin.
Painesville (14) A. Chemicals
                C. Textile
Cleveland (1 384) C. Prim. metal, Machin.
Lorain-Elyria (148) A. Primary metal
                C. Mot. vehicles,
                Fabricated metal
Akron (367)     A. Nondurable
Canton (174)    B. Prim. metal, Machin.
                C. Fabricated metal
Alliance (26)   B. Primary metal,
                Machinery
                C. Durable, Fabr. metal
Salem (13)      B. Fabricated metal,
                Machinery
                C. Prim. metal, Durable
Youngstown (298) A. Primary metal
East Liverpool (24) A. Durable
                B. Primary metal
Steubenville (36) A. Primary metal
New Philadelphia (13) C. Durable, Prim. metal,
                Machinery, Mining
Wooster (14)    C. *Education*, Nondur-
                able, *Retail*
Ashland (14)    B. Nondurable,
                Machinery
                C. Printing
Mansfield (44)  B. Machinery
                C. El. machin., Nondur-
                able, Fabricated metal
Mount Vernon (12) B. Machinery, Durable,
                Nondurable
Coshocton (12)  B. Durable
                C. Nondurable
Cambridge (15)  A. Durable
                C. *Medical*
Zanesville (41) B. Durable
                C. El. machinery
Newark (34)     B. Durable
                C. *Railroads*,
                El. machinery

| | |
|---|---|
| Lancaster (24) | A. Durable |
| Marietta (16) | C. Furn. & Lumber, Chemicals, Construction, *Retail* |
| Athens (12) | B. *Education*, |
| | C. Printing |
| Chillicothe (20) | A. Nondurable |
| | C. *Medical, Railroads* |
| Washington (11) | C. *Retail*, Food |
| Portsmouth (37) | B. Nondurable, Prim. metal, *Railroads* |
| Cincinnati (813) | C. Machinery |
| Hamilton (63) | B. Nondurable, Machinery |
| | C. Mot. vehicles, Fabric. metal |
| Middletown (34) | A. Primary metal |
| | B. Nondurable |
| Dayton (347) | B. Machinery |
| | C. *Administration*, El. machinery |
| Xenia (13) | B. *Administration* |
| | C. Machinery |
| Springfield (82) | B. Motor vehicles |
| | C. Machinery, Printing |
| Troy (11) | C. Furniture & Lumber, Machinery, El. machinery |
| Piqua (17) | B. Machinery |
| | C. Apparel |
| Sidney (11) | A. Machinery |
| | C. Fabricated metal |
| Bellefontaine (10) | A. *Railroads* |
| | C. *Retail* |
| Columbus (438) | C. *Administration* |
| Delaware (12) | C. *Education*, Durable |
| Marion (34) | B. Machinery |
| | C. *Railroads*, Prim. metal |
| Bucyrus (10) | B. Machinery |
| | C. Nondurable, *Railroads* |
| Tiffin (19) | B. Durable |
| | C. Machinery, El. mach. |
| Fostoria (14) | A. El. machinery |
| | C. Prim. metal, *Railroads* |
| Findlay (24) | B. Nondurable |
| | C. Machinery, *Retail* |
| Lima (88) | C. Electrical machinery |
| Van Wert (10) | B. Nondurable |
| Defiance (11) | C. Fabric. metal, Prim. metal, Food, *Retail* |
| Bowling Green (12) | B. *Education* |
| | C. *Eating places, Retail* |
| Fremont (17) | C. Fabric. metal, Apparel, Primary metal |
| Sandusky (29) | B. Machinery |
| | C. El. machinery, Nondurable, Primary metal |
| Toledo (364) | C. Mot. vehicles, Durable |

MICHIGAN

| | |
|---|---|
| Niles (13) | C. *Railroads*, Nondurable, Fabr. metal, Mot. vehicles, Machinery |
| Benton Harbor-St. Joseph (29) | B. Machinery |
| | C. Mot. vehicles, Prim. metal |
| Kalamazoo (83) | B. Nondurable |
| | C. Chemicals |
| Battle Creek-Springfield Place-Lakeview (62) | B. Food |
| | C. Machinery, *Medical* |
| Albion (10) | B. Primary metal, Fabricated metal |
| | C. *Education*, Machinery |
| Jackson (108) | B. Motor vehicles |
| Adrian (18) | B. Fabr. metal |
| | C. Mot. vehicles, Machin. |
| Monroe (21) | A. Nondurable |
| | B. Motor vehicles |
| Ann Arbor (48) | A. *Education* |
| | C. *Medical*, Mot. vehicles |
| Ypsilanti-Willow Run (30) | A. Motor vehicles |
| | B. *Education* |
| Detroit (2 659) | A. Motor vehicles |
| Port Huron (36) | B. Primary metal |
| | C. Motor vehicles, *Railroads* |
| Pontiac (93) | A. Motor vehicles |
| Flint (198) | A. Motor vehicles |
| Owosso (16) | B. Electr. machinery |
| | C. Mot. vehicles, Furn. & Lumber |
| Lansing (134) | A. Motor vehicles |
| | C. *Education, Administr.* |
| Grand Rapids (227) | C. Furn. & Lumber, Machinery, Mot. vehicles, Fabricated metal |
| Holland (16) | B. Furniture & Lumber |
| | C. Fabr. metal, Food |
| Muskegon (82) | A. Machinery |
| | C. Prim. metal, Motor vehicles |
| Mount Pleasant (11) | B. *Education*, Mining |
| | C. Motor vehicles |
| Midland (13) | A. Chemicals |
| Saginaw (106) | B. Mot. vehicles, Prim. metal |
| | C. Machinery |
| Bay City (88) | B. Motor vehicles |
| | C. Primary metal |
| Cadillac (10) | C. Furn. & Lumber, Nondurable, *Retail* |
| Traverse City (17) | B. *Medical* |
| | C. *Retail* |
| Alpena (13) | B. Durable |
| | C. *Retail*, Machinery |
| Sault Ste Marie (13) | B. Nondurable |
| | C. Chemicals, *Oth. transportation* |
| Escanaba (14) | B. *Railroads* |
| | C. Furniture & Lumber, Nondurable, *Retail* |
| Marinette, Wis.-Menominee, Mich. (25) | B. Furniture & Lumber, Nondurable |
| Marquette (17) | B. *Railroads* |
| | C. Furniture & Lumber, *Administration* |
| Ironwood (11) | A. Mining |
| | C. Furniture & Lumber |

INDIANA

| | |
|---|---|
| Goshen (13) | B. Furniture & Lumber, Nondurable |
| | C. Electrical machinery |

Elkhart (36)　　　C. Durable, *Railroads,* Mot. vehicles, El. machin., Chemicals

South Bend (168)　A. Motor vehicles
　　　　　　　　　　C. Machinery, Nondurable

La Porte (18)　　　B. Machinery
　　　　　　　　　　C. Motor vehicles

Michigan City (28)　C. Fabr. metal, Transp. equipment, Machinery, *Railroads*

Valparaiso (12)　　C. *Education,* Durable, Prim. metal, El. machin.

Fort Wayne (140)　B. Electrical machinery
　　　　　　　　　　C. Mot. vehicles, Machinery

Huntington (15)　　C. Furniture & Lumber, *Railroads,* El. machinery

Wabash (11)　　　B. Nondurable
　　　　　　　　　　C. Furniture & Lumber, Durable

Peru (13)　　　　A. *Railroads*
　　　　　　　　　　C. Electrical machinery, Furniture & Lumber

Logansport (21)　B. *Railroads*
　　　　　　　　　　C. Mot. vehicles, Fabric. metal, *Retail*

Marion (30)　　　B. Primary metal
　　　　　　　　　　C. Nondurable, Electrical machinery

Kokomo (39)　　　B. Primary metal, Electrical machinery
　　　　　　　　　　C. Mot. vehicles, Fabr. metal

Lafayette-
West Lafayette (47)　B. *Education*
　　　　　　　　　　C. Primary metal

Frankfort (15)　　B. *Railroads*
　　　　　　　　　　C. Fabr. metal

Elwood (11)　　　B. Electric machinery, Fabricated metal

Anderson (47)　　A. Electrical machinery
　　　　　　　　　　C. Fabricated metal

Muncie (90)　　　B. Motor vehicles
　　　　　　　　　　C. Durable

New Castle (18)　A. Motor vehicles
　　　　　　　　　　C. Machinery

Richmond (40)　　A. Machinery

Connersville (16)　A. Machinery
　　　　　　　　　　B. Motor vehicles

Shelbyville (12)　B. Furniture & Lumber
　　　　　　　　　　C. Nondurable, Machin., Electrical machinery

Indianapolis (502)

Crawfordsville (13)　B. Printing
　　　　　　　　　　C. Prim. metal, *Retail*

Terre Haute (78)　C. Food, *Railroads*

Bloomington (28)　A. *Education*
　　　　　　　　　　B. Electrical machinery

Columbus (18)　　B. Machinery, El. machin.
　　　　　　　　　　C. Furniture & Lumber

Bedford (13)　　　B. Prim. metal, Durable
　　　　　　　　　　C. *Administration*

Washington (11)　B. *Railroads*
　　　　　　　　　　C. Nondurable

Vincennes (19)　　C. *Retail,* Nondurable

Evansville (138)　B. Machinery
　　　　　　　　　　C. Food

ILLINOIS

Waukegan (39)　　C. Prim. metal, Machin., Chemicals, *Administrat.,*Fabricated metal

Chicago (4,921)　C. Primary metal

Elgin (44)　　　　B. Durable
　　　　　　　　　　C. *Medical,* El. machin.

Aurora (51)　　　B. Machinery
　　　　　　　　　　C. *Railroads,* Furniture & Lumber

Joliet (52)　　　C. Prim. metal, *Railroads,* Nondurable

De Kalb (12)　　C. Electric machinery, Durable, *Education,* Primary metal

Rockford (122)　A. Machinery
　　　　　　　　　　B. Fabr. metal

Freeport (22)　　B. Electrical machinery
　　　　　　　　　　C. *Railroads, Retail,* Chemicals

Sterling (12)　　B. Primary metal, Fabr. metal

Dixon (12)　　　C. *Utilities,* Prim. metal, Nondurable

Kewanee (17)　　A. Machinery
　　　　　　　　　　B. Fabricated metal
　　　　　　　　　　C. *Medical*

La Salle (12)　　A. Durable
　　　　　　　　　　C. Primary metal

Ottawa (17)　　　A. Durable

Streator (16)　　A. Durable
　　　　　　　　　　C. Motor vehicles
　　　　　　　　　　B. Fabricated metal

Kankakee (26)　　C. *Medical,* Furniture & Lumber

Bloomington (34)　C. *Finance, Retail, Railroads*

Peoria (155)　　　A. Machinery
　　　　　　　　　　C. Food

Pekin (22)　　　B. Food, Machinery

Galesburg (31)　B. *Railroads*
　　　　　　　　　　C. *Retail*

Monmouth (10)　C. *Retail,* Durable, *Finance*

Canton (12)　　　A. Machinery
　　　　　　　　　　C. Mining

Macomb (11)　　C. Durable, *Retail, Education,* Fabr. metal

Quincy (41)　　　C. Machinery

Jacksonville (20)　B. *Medical, Education*
　　　　　　　　　　C. *Retail*

Springfield (97)　C. *Administration,* Machinery

Lincoln (14)　　　B. Durable
　　　　　　　　　　C. *Medical, Retail*

Decatur (74)　　C. *Railroads,* Food

Champaign-Urbana (62)　A. *Education*

Danville (38)　　C. *Railroads, Medical, Retail*

Mattoon (18)　　B. *Railroads*
　　　　　　　　　　C. Nondurable, *Retail*

Alton (33)　　　B. Durable, Primary metal
　　　　　　　　　　C. Nondurable

| | |
|---|---|
| Wood River (10) | A. Nondurable<br>C. Primary metal, Fabric. metal |
| Collinsville (12) | C. Food, Mining, Apparel |
| Centralia (14) | B. *Railroads*<br>C. *Retail* |
| Mount Vernon (16) | C. *Retail*, Mining |
| West Frankfort (11) | A. Mining |
| Carbondale (11) | B. *Education, Railroads*<br>C. Construction |
| Marion (10) | B. Mining<br>C. El. machinery, *Retail* |
| Harrisburg (11) | A. Mining<br>C. *Retail* |
| Cairo (12) | C. *Retail, Wholesale, Households* |

WISCONSIN

| | |
|---|---|
| Ashland (11) | B. *Railroads* |
| Chippewa Falls (11) | B. Nondurable<br>C. *Medical* |
| Eau Claire (36) | B. Nondurable |
| Marshfield (12) | B. Furniture & Lumber<br>C. Nondurable, *Wholesale, Retail, Medical* |
| Wausau (30) | C. Furniture & Lumber, Nondurable, *Finance* |
| Wisconsin Rapids (13) | A. Nondurable<br>C. Fabricated metal |
| Stevens Point (17) | C. Nondurable, *Finance*, Furniture & Lumber, *Railroads*, Durable |
| Green Bay (53) | B. Nondurable<br>C. *Railroads*, Food, *Retail, Wholesale* |
| Two Rivers (10) | A. Furniture & Lumber, Fabricated metal |
| Manitowoc (28) | B. Fabricated metal, Transport. equipment |
| Sheboygan (42) | B. Fabricated metal<br>C. Furniture & Lumber, Nondurable, Durable |
| Appleton (34) | B. Nondurable |
| Menasha-Neenah (25) | A. Nondurable<br>C. Printing |
| Oshkosh (41) | B. Furniture & Lumber<br>C. Motor vehicles |
| Fond du Lac (30) | B. Machinery<br>C. *Railroads, Retail* |
| La Crosse (48) | C. Machinery, Motor vehicles |
| Beaver Dam (12) | B. Fabricated metal, Nondurable<br>C. Food |
| Watertown (12) | C. Nondurable, Machin., Fabricated metal |
| Madison (110) | B. *Education*<br>C. *Administration*, Food |
| Janesville (25) | B. Motor vehicles, Durable |
| Beloit (30) | A. Machinery<br>C. Nondurable, Electric machinery |
| Waukesha (21) | B. Machinery, Primary metal |
| Milwaukee (830) | B. Machinery |

| | |
|---|---|
| Racine (77) | A. Machinery<br>C. Electrical machinery, Fabricated metal, Primary metal |
| Kenosha (54) | A. Motor vehicles<br>B. Furniture & Lumber<br>C. Primary metal |

MINNESOTA

| | |
|---|---|
| Virginia (12) | A. Mining<br>C. *Railroads*, Apparel |
| Hibbing (16) | A. Mining |
| Bemidji (10) | C. *Retail, Education,* Furniture & Lumber |
| Fergus Falls (13) | B. *Medical*<br>C. *Retail, Utilities* |
| Brainerd (13) | A. Railroads |
| Duluth-Superior (143) | B. Railroads<br>C. Primary metal, *Wholesale* |
| St. Cloud (28) | C. *Railroads, Medical, Retail* |
| Minneapolis-St. Paul (985) | C. *Wholesale* |
| Red Wing (11) | B. Nondurable, Durable |
| Faribault (16) | C. *Education, Medical,* Food |
| Owatonna (10) | C Durable, *Finance, Retail,* Food |
| Mankato (19) | C. *Retail, Wholesale,* Construction, *Education* |
| Albert Lea (14) | B. Food<br>C. Fabr. metal, *Retail* |
| Austin (23) | A. Food |
| Rochester (30) | A. *Medical* |
| Winona (24) | C. Food, *Education* |

IOWA

| | |
|---|---|
| Mason City (28) | B. Food<br>C. Durable, *Retail, Wholesale* |
| Charles City (10) | A. Machinery |
| Waterloo (84) | A. Machinery<br>B. Food |
| Dubuque (50) | C. Food, Furn. & Lumb. Machinery |
| Cedar Rapids (78) | B. Food<br>C. Machinery, *Retail* |
| Iowa City (27) | A. *Education*<br>B. *Medical* |
| Clinton (30) | B. Food<br>C. Chemicals, *Railroads* |
| Davenport, Iowa-Rock Island, Ill.-Moline, Ill. (195) | A. Machinery |
| Muscatine (19) | B. Durable<br>C. Furn. & Lumber, Food |
| Burlington (31) | C. *Wholesale, Railroads, Retail,* Machinery |
| Fort Madison (15) | A. Durable<br>B. *Railroads*<br>C. Chemicals |
| Keokuk (16) | B. Nondurable<br>C. Food, Primary metal |

| Ottumwa (35) | A. Food |
| | C. Machinery, *Railroads* |
| Oskaloosa (11) | C. *Retail* |
| Newton (12) | A. Machinery |
| Des Moines (200) | C. *Finance, Retail* |
| Marshalltown (20) | C. Machinery, Fabricated |
| | metal, *Retail* |
| Ames (23) | A. *Education* |
| | C. Construction |
| Boone (12) | B. *Railroads* |
| | C. *Retail*, Construction |
| Fort Dodge (25) | C. Food, *Retail*, Durable, |
| | *Wholesale* |
| Sioux City (90) | B. Food |
| | C. *Wholesale, Retail* |

MISSOURI

| Kirksville (11) | C. Nondurable, *Medical*, |
| | *Retail, Education* |
| Hannibal (20) | B. Nondurable |
| | C. *Railroads, Retail* |
| Moberly (13) | B. *Railroads*, Nondurable |
| Mexico (12) | A. Durable |
| | C. Nondurable |
| Columbia (32) | A. *Education* |
| Fulton (10) | B. Nondurable, *Education, Medical* |
| Jefferson City (25) | B. *Administration* |
| | C. Nondurable, Construction |
| Sedalia (20) | B. *Railroads* |
| | C. *Retail, Wholesale* |
| St. Joseph (82) | B. Food |
| | C. *Wholesale, Retail* |
| Kansas City (698) | C. *Retail* |
| Carthage (11) | C. *Retail*, Apparel |
| Joplin (39) | C. *Retail, Wholesale* |
| Springfield (76) | C. *Railroads, Retail, Wholesale* |
| St. Charles (14) | B. Transp. equipment, Nondurable |
| St. Louis (1,400) | |
| Cape Girardeau (22) | B. Nondurable |
| | C. *Retail* |
| Sikeston (12) | B. Nondurable |
| | C. *Retail*, Construction |
| Poplar Bluff (15) | C. *Railroads, Retail*, Construction, Nondurable |

NORTH DAKOTA

| Fargo, N. D.-Moorhead, Minn. (53) | C. *Wholesale, Retail* |
| Grand Forks (27) | C. *Railroads, Retail, Wholesale*, Food, Education |
| Jamestown (11) | B. *Railroads* |
| | C. *Medical, Retail* |
| Bismarck (19) | B. *Administration* |
| | C. *Medical, Retail, Wholesale*, Construction |
| Minot (21) | B. *Railroads* |
| | C. *Retail, Wholesale, Medical* |

SOUTH DAKOTA

| Sioux Falls (71) | B. Food |
| | C. *Wholesale, Retail* |
| Mitchell (12) | B. *Retail* |
| | C. *Wholesale, Medical* |
| Huron (13) | C. *Railroads, Retail*, Construction, Food, *Wholesale* |
| Watertown (13) | B. *Retail* |
| | C. *Wholesale*, Food |
| Aberdeen (21) | B. *Retail* |
| | C. *Wholesale, Railroads* |
| Rapid City (25) | B. Construction, *Retail, Administration* |

NEBRASKA

| Beatrice (12) | C. *Retail*, Machinery, Furniture & Lumber |
| Lincoln (100) | C. *Retail, Education, Administration, Railroads* |
| Omaha (310) | B. Food |
| | C. *Railroads, Finance, Wholesale* |
| Fremont (15) | B. Food |
| | C. *Retail, Wholesale*, Construction |
| Norfolk (11) | B. *Retail* |
| | C. *Wholesale, Railroads, Medical* |
| Grand Island (23) | C. *Railroads*, Construction, *Retail, Wholesale* |
| Hastings (20) | C. *Retail, Wholesale, Medical* |
| Kearney (12) | C. *Retail*, Construction, *Medical, Education* |
| North Platte (15) | A. *Railroads* |
| | C. *Retail* |
| Scottsbluff (13) | B. *Retail* |
| | C. Food, *Wholesale*, Construction |

KANSAS

| Coffeyville (17) | C. Nondurable, *Retail, Railroads* |
| Independence (11) | C. Nondurable, *Retail*, Durable |
| Parsons (15) | A. *Railroads* |
| | C. *Retail* |
| Pittsburg (19) | B. *Railroads* |
| | C. *Retail, Education* |
| Fort Scott (10) | B. *Railroads* |
| | C. *Finance, Retail* |
| Chanute (10) | C. *Railroads, Retail*, Durable |
| El Dorado (11) | B. Nondurable |
| | C. Mining, *Retail*, Construction |
| Winfield (10) | C. *Education, Retail, Medical* |
| Arkansas City (13) | B. *Railroads*, Food |
| | C. *Retail* |
| Wichita (194) | B. Transp. equipment |
| | C. *Retail* |

Newton (12)
A. *Railroads*
C. *Retail*

Emporia (16)
B. *Railroads*
C. *Education, Retail*

Ottawa (10)
C. *Retail*, Construction

Lawrence (23)
A. *Education*

Topeka (89)
B. *Railroads*
C. *Medical, Administration, Food*

Leavenworth (21)
B. *Administration*
C. *Medical*

Atchison (13)
C. *Railroads, Wholesale,* Transp. equipm., Food, *Retail*

Manhattan (19)
A. *Education*
C. *Retail, Administration*

Junction City (13)
B. *Administration, Retail,*
C. *Eating places*

Salina (26)
B. *Retail*
C. *Wholesale,* Food Construction

Hutchinson (34)
C. *Retail,* Construction, *Wholesale,* Food,

Great Bend (13)
B. Mining
C. Construction, *Retail*

Dodge City (11)
B. *Retail*
C. *Railroads, Wholesale,* Construction

Garden City (11)
B. *Retail*
C. Construction

VIRGINIA

Bristol, Va.-
Bristol, Tenn. (33)
C. Apparel, Textile, Furniture & Lumber

Martinsville (17)
B. Furniture & Lumber, Textile, Chemicals

Danville (35)
A. Textile

Roanoke (107)
B. *Railroads*

Lynchburg (48)
B. Nondurable
C. Textile, Apparel

Staunton (20)
B. Textile
C. Apparel, *Education, Medical*

Waynesboro (12)
A. Chemicals
C. Textile

Charlottesville (26)
C. *Education,* Textile, Medical, Households, Retail

Harrisonburg (11)
C. *Retail, Wholesale*

Winchester (14)
C. Textile, *Retail, Households*

Fredericksburg (12)
B. Chemicals
C. Apparel, Construction

Richmond (258)
C. Nondurable

Hopewell (10)
A. Chemicals
B. Textile
C. Nondurable

Petersburg (35)
B. Nondurable
C. *Households,* Chemicals, Durable, *Administration*

Newport News-Newsome Park-Hilton Park-Riverview (72)
B. Transp. equipment
C. *Administration, Households*

Norfolk-Portsmouth (385)
B. *Administration*
C. Transp. equipment

Suffolk (12)
B. *Wholesale*
C. *Retail,* Furniture & Lumber

NORTH CAROLINA

Elizabeth City (13)
C. Furniture & Lumber, Households, Textile, Retail, Administration

Greenville (17)
B. Households
C. Retail, Construction, Education

New Bern (16)
B. Administration
C. Households, Furniture & Lumber, *Retail*

Kinston (18)
C. Households, Retail, Apparel, Construction

Goldsboro (21)
B. Furniture & Lumber
C. Households, Retail

Wilson (22)
C. Households, Construction, *Retail*

Rocky Mount (28)
B. Railroads
C. Textile, *Households, Retail*

Henderson (11)
B. Textile
C. Households, Retail

Raleigh (69)
C. Administration, Education, Households

Durham (73)
B. Nondurable, Textile
C. Education, Medical, Construction

Burlington (25)
A. Textile

Reidsville (12)
A. Nondurable
C. Textile

Greensboro (83)
B. Textile
C. Households

Winston-Salem (92)
A. Nondurable
C. Textile, *Households*

High Point (40)
A. Textile
B. Furniture & Lumber

Thomasville (11)
A. Furniture & Lumber, Textile

Lexington (14)
A. Textile
B. Furniture & Lumber

Salisbury (20)
B. Textile, *Railroads*
C. *Retail*

Kannapolis (28)
A. Textile

Concord (16)
A. Textile
C. *Households*

Albemarle (12)
A. Textile

Statesville (17)
B. Textile, Furniture & Lumber

Hickory (15)
A. Textile
B. Furniture & Lumber

Asheville (58)
C. *Households, Medical, Retail*

Shelby (16)
A. Textile
C. *Households*

Gastonia (23)
A. Textile
C. *Households*

Charlotte (141)
C. Textile, *Households, Wholesale*

Monroe (10)
B. Textile
C. Construction, *Households*

Sanford (10)
C. Textile, Machinery, Furniture & Lumber, *Households*

Fayetteville (35)    C. *Households, Retail, Administration*

Wilmington (45)    C. *Railroads, Households*

SOUTH CAROLINA

Rock Hill (25)    A. Textile
C. Chemicals, *Education*

Spartanburg (37)    B. Textile
C. *Households*

Greenville (168)    A. Textile

Anderson (20)    A. Textile
C. *Households, Retail*

Greenwood (14)    A. Textile
C. *Households*

Columbia (121)    C. *Households,* Textile, Construction, *Administration*

Orangeburg (15)    B. *Households*
C. Furniture & Lumber, *Education,* Textile, *Retail*

Sumter (20)    B. Furniture & Lumber
C. *Households, Retail*

Florence (23)    B. *Railroads*
C. *Households, Retail*

Charleston (120)    B. Transport. equipment
C. *Households*

GEORGIA

Dalton (16)    A. Textile
B. Apparel

Rome (30)    B. Textile
C. *Households*

Gainesville (12)    C. Textile, *Households,* Food, *Retail, Wholesale*

Marietta (21)    C. *Administration,* Construction, Textile, *Retail*

Atlanta (508)    C. *Households*

Athens (28)    B. *Education, Households*
C. Textile

Augusta (88)    B. Textile
C. *Households*

Midway-Hardwick (15)    A. *Medical*

Dublin (10)    B. *Households*
C. Furniture & Lumber, *Retail, Medical*

Macon (93)    C. *Households,* Textile, *Administration*

Griffin (14)    A. Textile
C. *Households*

La Grange (25)    A. Textile
C. *Households*

Columbus (118)    A. Textile
C. *Households*

Americus (11)    B. *Households,* Furniture & Lumber
C. *Railroads*

Albany (31)    B. *Households*
C. Textile, *Wholesale, Retail*

Moultrie (12)    B. *Households*
C. Food, *Retail*

Thomasville (14)    B. *Households*
C. Furniture & Lumber, Food

Valdosta (20)    B. *Households*
C. Furniture & Lumber, *Retail*

Waycross (19)    B. *Railroads*
C. *Households,* Nondurable

Brunswick (18)    C. Chemicals, *Households,* Food

Savannah (128)    B. Nondurable
C. *Households, Railroads*

FLORIDA

Pensacola-Warrington-Brownsville-Brent-Goulding (77)    B. *Administration*
C. *Households*

Panama City (26)    C. Nondurable, *Administration,* Households, Construction, *Retail*

Tallahassee (27)    B. *Education, Administration*
C. *Households,* Construction

Jacksonville (243)    C. *Households, Administration,* Railroads, *Wholesale*

St. Augustine (14)    B. *Railroads*
C. *Households, Education, Hotels*

Gainesville (27)    A. *Education*
C. *Households*

Ocala (12)    C. *Retail, Households,* Furniture & Lumber

Daytona Beach (30)    C. *Households,* Construction, *Retail, Eating places*

Sanford (12)    B. *Wholesale*
C. *Railroads, Households*

Orlando (73)    C. *Households, Retail,* Construction, *Wholesale*

Lakeland (31)    C. *Households, Retail*

Tampa (179)    C. Nondurable, *Wholesale,* Construction

Clearwater (16)    C. *Households,* Construction, *Retail*

St. Petersburg (115)    C. Construction, *Retail, Households*

Bradenton (14)    C. *Retail, Households,* Furniture & Lumber, Construction

Sarasota (19)    B. *Households*
C. Construction, *Entertainment, Retail*

Fort Pierce (14)    B. *Wholesale*
C. *Retail,* Construction

West Palm Beach (43)    B. *Households*
C. *Retail,* Construction

Lake Worth (12)    B. Construction
C. *Retail*

Fort Myers (13)    C. *Retail, Households,* Construction

Fort Lauderdale (36)    B. Construction
C. *Retail, Households*

Hollywood (14)
B. *Hotels*
C. Construction, *Finance*

Miami (459)
C. Construction, *Oth. transport, Retail*

Key West (26)
A. *Administration*
C. *Eating places*

ALABAMA

Florence (24)
C. Prim. metal, Construction, Textile, *Households, Utilities, Retail*

Sheffield (11)
B. Prim. metal
C. Chemicals, *Railroads,* Construction

Decatur (20)
B. Textile
C. *Retail, Households*

Huntsville (16)
C. *Administration, Retail, Households*

Gadsden (94)
A. Prim. metal
B. Textile
C. Nondurable

Anniston (31)
B. Prim. metal, Textile
C. *Households,* Administr.

Talladega (13)
B. Textile
C. *Education,* Primary metal, *Households*

Birmingham (445)
B. Prim. metal
C. *Households,* Mining

Tuscaloosa (46)
B. *Education*
C. Nondurable, *Households, Medical*

Selma (23)
B. *Households*

Montgomery (109)
B. *Households*
C. *Administration*

Auburn (13)
A. *Education*
B. *Households*

Opelika (12)
A. Textile
B. *Households*

Dothan (22)
B. *Households*
C. *Retail, Wholesale*

Mobile (183)
C. *Administration, Households, Oth. transport.* Nondurable

MISSISSIPPI

Tupelo (12)
C. *Households, Wholesale, Retail,* Apparel

Clarksdale (17)
B. *Households*
C. *Retail,* Construction

Greenville (30)
B. *Households*
C. Construction, *Retail,* Furniture & Lumber

Greenwood (18)
B. *Households*
C. *Retail, Wholesale,* Construction

Columbus (17)
B. *Households*
C. Furniture & Lumber, *Education*

Meridian (42)
C. *Households, Railroads,* Furn. & Lumber, *Retail*

Jackson (100)
C. *Households,* Construction

Vicksburg (28)
B. *Households,* Construction
C. Furniture & Lumber, *Administration*

Natchez (23)
B. Nondurable, *Households*
C. Furniture & Lumber, Construction

McComb (10)
B. *Railroads*
C. *Retail, Households*

Laurel (25)
B. Nondurable
C. Furniture & Lumber, *Households*

Hattiesburg (29)
C. *Households,* Chemicals, *Retail*

Gulfport (23)
C. *Medical, Administration,* Construction, *Households*

Biloxi (37)
B. Food, *Administration*

Pascagoula (11)
B. Transp. equipment
C. Construction

LOUISIANA

Bastrop (13)
A. Nondurable

Monroe - West
Monroe (49)
C. *Households, Retail, Wholesale*

Ruston (10)
B. *Education, Households*

Shreveport (150)
C. *Households, Retail,* Mining, Construction

Alexandria (35)
C. *Households, Retail,* Furniture & Lumber

Lake Charles (41)
C. Nondurable, Construction, *Households, Retail*

Crowley (13)
B. Food
C. *Households, Retail,* Construction

Opelousas (12)
C. *Retail, Households, Education*

Lafayette (34)
C. *Households, Railroads,* Construction, *Education*

New Iberia (16)
B. Mining
C. *Households,* Construct., *Wholesale*

Baton Rouge (139)
B. Nondurable
C. Construction, *Households, Education,* Chemicals

Houma (12)
B. Mining
C. Food

New Orleans (660)
C. *Oth. transportation, Wholesale*

Bogalusa (18)
A. Nondurable
C. Furniture & Lumber, Construction, *Households*

KENTUCKY

Paducah (33)
B. *Railroads*

Henderson (17)
C. Furniture & Lumber

Owensboro (34)
B. El. machinery
C. Food

Madisonville (11)
A. Mining

Hopkinsville (13)
C. *Retail, Households*

Bowling Green (18)
C. Apparel, Construction *Retail, Households*

Louisville (473)

Frankfort (12)
B. *Administration,* Food
C. Construction

Lexington (101)   C. *Medical, Retail,* Construction, *Households, Education*

Richmond (10)   B. *Administ., Education,* C. *Households,* El. machin.

Middlesborough (14)   B. Mining C. *Retail*

TENNESSEE

Kingsport (20)   A. Chemicals C. Textile, Printing

Elizabethton (11)   A. Chemicals C. Construction, *Retail*

Johnson City (28)   C. Textile, *Retail, Medical*

Morristown (13)   B. Furniture & Lumber C. Textile, Chemicals

Knoxville (148)   C. Textile, Construction, Retail

Oak Ridge (30)   A. Chemicals B. Construction C. *Administration*

Cleveland (13)   B. Textile, Fabr. metal

Chattanooga (168)   B. Textile

Murfreesboro (13)   C. *Households, Education, Medical*

Columbia (11)   C. Chemicals, *Households*

Nashville (259)

Clarksville (16)   B. Nondurable C. *Households*

Dyersburg (11)   B. Textile C. *Households, Retail*

Jackson (30)   C. *Railroads, Households, Retail*

Memphis (406)   C. *Households, Wholesale, Retail*

ARKANSAS

Blytheville (16)   B. *Retail* C. *Households,* Construction

Jonesboro (16)   C. *Retail, Nondurable, Wholesale*

Fayetteville (17)   B. *Education* C. *Medical*

Fort Smith (56)   C. *Retail, Wholesale,* Furniture & Lumber

Hot Springs (29)   C. *Hotels, Retail, Personal, Medical, Eating*

Little Rock (154)   C. *Railroads*

Helena (11)   B. Furniture & Lumber, *Households* C. *Retail*

Pine Bluff (37)   B. *Railroads* C. *Households, Retail*

Camden (11)   B. Nondurable, *Households* C. Furniture & Lumber, *Retail*

El Dorado (23)   B. Nondurable C. *Households,* Chemicals, Mining

OKLAHOMA

Durant (11)   B. *Retail* C. *Education*

Ardmore (18)   C. *Retail,* Mining, *Households*

Duncan (15)   A. Mining C. *Retail,* Construction

Lawton (35)   C. Construction, *Retail, Administration*

Ada (16)   C. *Retail,* Mining, Durable

McAlester (18)   B. *Administration* C. *Retail, Apparel*

Seminole (12)   A. Mining C. Construction

Shawnee (23)   C. *Retail, Administration,* Mining

Norman (27)   A. *Education* C. *Medical*

Chickasha (16)   C. *Retail,* Mining

El Reno (11)   A. *Railroads* C. *Retail, Administration*

Oklahoma City (275)   B. *Administration* C. *Retail,* Construction

Okmulgee (18)   B. Durable C. *Education, Retail*

Muskogee (37)   C. *Retail, Administration*

Tulsa (206)   C. Mining, Nondurable

Sapulpa (13)   B. Durable C. Mining, *Railroads*

Stillwater (20)   A. *Education* C. Construction

Guthrie (10)   C. Furniture & Lumber, *Retail*

Enid (36)   C. *Retail, Wholesale*

Ponca City (20)   A. Nondurable C. *Retail*

Bartlesville (19)   A. Mining C. Construction

Miami (12)   B. Nondurable C. *Retail,* Mining

TEXAS

Orange (21)   B. Chemicals C. *Fabricated metal,* Construction

Port Arthur (82)   A. Nondurable

Beaumont (94)   B. Nondurable C. Construction, *Households*

Baytown (23)   A. Nondurable

Houston (701)   C. Construction, Nondurable

Texas City (17)   B. Nondurable, Chemicals C. Construction

Galveston (72)   B. *Oth. transportation*

Lufkin (15)   B. Primary metal C. *Households,* Furniture & Lumber, Nondurable, *Retail*

Nacogdoches (12)   B. Furniture & Lumber C. *Education, Households, Retail*

Palestine (13)   A. *Railroads* C. *Households, Retail*

Tyler (39)   C. *Households, Retail, Railroads*

Longview (25)   C. *Households,* Construct., *Retail,* Machinery

Marshall (22) — C. *Railroads, Households, Education, Retail*

Texarkana, Tex.-Texarkana, Ark. (41) — B. *Administration* / C. *Retail, Households, Railroads*

Paris (22) — C. *Retail, Households*

Denison (18) — B. *Railroads*

Sherman (20) — C. Food, *Retail*

Gainesville (11) — C. Mining, *Retail*

Denton (21) — A. *Education* / C. *Retail*, Construction

McKinney (11) — B. *Medical* / C. Textile, *Retail*

Greenville (15) — C. *Retail*, Apparel, *Households*, Construct.

Fort Worth (316) — B. *Transp. equipment* / C. Food

Dallas (539) — C. *Retail, Finance,* Construction, *Wholesale*

Garland (11) — C. Food, Construction, *Retail, Wholesale*

Terrell (12) — B. *Medical* / C. Construction, *Households*

Waxahachie (11) — C. *Retail, Apparel,* Construction

Cleburne (13) — A. *Railroads* / C. *Retail*

Corsicana (19) — C. *Households, Retail,* Textile

Waco (93) — C. *Retail, Wholesale*

Temple (25) — B. *Medical* / C. *Railroads, Retail*

Bryan (18) — B. *Education* / C. *Households,* Construction, *Retail*

Austin (136) — C. *Education, Administration,* Construction, *Households*

New Braunfels (12) — A. Textile / C. Construction

San Antonio (450) — B. *Administration* / C. *Retail,* Construction

Victoria (16) — B. Construction / C. *Retail, Mining, Households*

Corpus Christi (133) — C. Construction, *Administration, Retail*

Alice (16) — B. Mining, Construction

Kingsville (17) — B. *Railroads* / C. *Education,* Construct., Chemicals

Brownsville (36) — C. Construction, *Retail, Households*

San Benito (13) — B. *Wholesale* / C. Construction, Food

Harlingen (23) — C. *Retail,* Construction, *Wholesale,* Food

Mercedes (10) — B. *Wholesale* / C. Retail, *Households*

Edinburg (12) — B. *Wholesale* / C. *Retail, Education*

McAllen (20) — C. *Retail, Wholesale,* Food, Construction

Mission (11) — C. *Wholesale, Households, Retail, Mining,* Food

Laredo (56) — C. *Retail, Households,* Construction, *Wholesale, Administration*

Del Rio (14) — C. *Retail, Households, Railroads,* Construction

San Angelo (59) — B. Construction / C. *Retail*

Brownwood (20) — C. *Retail, Railroads*

Abilene (46) — C. *Retail,* Construction, Mining, *Wholesale*

Sweetwater (14) — C. *Retail, Durable,* Mining

Snyder (12) — A. Mining / C. Construction

Big Spring (17) — C. Mining, *Retail, Railroads,* Construction

Midland (22) — A. Mining / B. Construction

Odessa (29) — A. Mining / C. Construction, *Wholesale*

Lamesa (29) — B. Construction / C. *Retail,* Mining

Lubbock (101) — C. Construction, *Retail, Wholesale*

Plainview (14) — B. *Retail* / C. Construction

Vernon (13) — B. *Retail* / C. Construction

Wichita Falls (98) — B. Mining / C. *Retail,* Construction

Amarillo (74) — C. Construction, *Wholesale, Retail, Railroads*

Borger (18) — B. Mining / C. Nondurable, Chemicals, Construction

Pampa (17) — B. Mining / C. Construction, *Retail,* Chemicals

El Paso (137) — C. *Railroads, Retail, Administration*

NEW MEXICO

Santa Fe (28) — B. Construction, *Administration* / C. *Retail*

Albuquerque (146) — B. Construction / C. *Administration, Retail*

Clovis (17) — B. *Railroads* / C. *Retail,* Construction

Roswell (26) — C. *Retail,* Construction, *Administration*

Hobbs (14) — A. Mining / C. Construction, *Retail*

Carlsbad (18) — A. Mining / C. Construction

Las Cruces (12) — B. Construction / C. *Administration, Educat., Retail, Households*

ARIZONA

Tucson-Amphitheater (58) — C. *Railroads, Retail, Education,* Construction

Phoenix (216)    C. Construction, *Retail, Wholesale*

NEVADA

Las Vegas (25)    B. *Entertainment, Hotels*
   C. *Eating places*
Reno (32)    C. *Entertainment,* Construction

UTAH

Logan (17)    B. *Education*
   C. *Retail,* Construction
Ogden (83)    A. *Administration*
   B. *Railroads*
   C. Food
Salt Lake City (227)    C. *Wholesale, Retail*
Provo (29)    B. *Prim. metal, Education*
   C. *Retail,* Construction

COLORADO

Fort Collins (15)    B. *Education*
   C. *Retail,* Construction
Greeley (20)    C. *Education, Retail,* Construction
Boulder (20)    A. *Education*
Denver (499)    C. *Administration*
Colorado Springs (45)    C. *Retail,* Construction
Pueblo (73)    A. Primary metal
   C. *Railroads, Medical*
Trinidad (12)    B. Mining
   C. *Railroads, Retail*
Grand Junction (15)    B. *Railroads*
   C. *Retail,* Construction

WYOMING

Sheridan (12)    C. *Medical, Retail, Railroads,* Construction
Caspar (24)    B. Nondurable
   C. *Mining,* Construction
Rock Springs (11)    A. *Mining*
   B. *Railroads*
Laramie (16)    B. *Education, Railroads*
Cheyenne (32)    B. *Railroads, Administration*

MONTANA

Great Falls (39)    C. *Prim. metal, Railroads, Retail*
Missoula (22)    C. *Railroads, Education, Retail,* Furn. & Lumber
Helena (18)    B. *Administration*
   C. Construction, *Finance*
Anaconda (11)    A. Primary metal
   C. *Railroads*
Butte (33)    A. Mining
Bozeman (11)    B. *Education*
   C. *Retail*
Billings (32)    C. *Wholesale, Retail,* Construction, *Administration*

IDAHO

Coeur d'Alene (12)    B. Furniture & Lumber
   C. Construction, Prim. metal
Moscow (11)    A. *Education*
Lewiston (13)    B. Furniture & Lumber
   C. *Retail*
Caldwell (10)    C. *Retail,* Construction, *Wholesale,* Food
Nampa (16)    B. *Railroads*
   C. Food, *Retail,* Construction
Boise City (34)    C. *Administration, Retail,* Construction, *Finance*
Twin Falls (18)    B. *Retail*
   C. *Wholesale,* Construct.
Idaho Falls (19)    B. *Retail*
   C. Construct., *Wholesale*
Pocatello (26)    A. *Railroads*

WASHINGTON

Walla Walla (24)    C. Construction, *Retail, Medical*
Pullman (12)    A. *Education*
Spokane (176)    C. *Retail, Railroads*
Pasco-Kennewick (20)    A. Construction
   C. *Railroads*
Richland (22)    A. Chemicals
   C. Construction
Yakima (38)    C. *Retail, Wholesale,* Construction
Wenatchee (13)    C. *Retail,* Construction, *Wholesale*
Longview (20)    A. Furniture & Lumber
   B. Nondurable
Aberdeen-Hoquiam (31)    A. Furniture & Lumber
   C. Nondurable
Olympia (16)    A. *Administration*
Tacoma (168)    C. Furniture & Lumber, *Administration*
Puyallup (10)    B. Furniture & Lumber
Renton (15)    A. Transp. equipment
Bremerton (28)    A. Transp. equipment
Seattle (622)    C. Transp. equipment
Everett (34)    B. Furniture & Lumber
   C. Nondurable
Port Angeles (11)    A. Nondurable
   B. Furniture & Lumber
Bellingham (34)    B. Furniture & Lumber
   C. *Retail*

OREGON

Medord (17)    B. Furniture & Lumber
   C. *Retail, Wholesale*
Klamath Falls (16)    B. Furniture & Lumber,
   C. *Railroads, Retail*
Bend (11)    A. Furniture & Lumber
Eugene-Springfield (47)    B. Furniture & Lumber
   C. *Retail, Education*
Corvallis (16)    A. *Education*
   C. *Hotels*
Albany (10)    B. Furniture & Lumber
   C. *Retail*
Salem (43)    B. *Administration*
   C. *Retail,* Construction, *Medical*

Portland (513)
C. *Wholesale*

Pendleton (12)
C. Furniture & Lumber, *Medical,* Construction, *Retail, Railroads*

Astoria (12)
B. Food
C. *Administration, Retail, Oth. transportation*

CALIFORNIA

El Centro (13)
C. *Wholesale, Retail, Administration*

Brawley (12)
C. *Wholesale, Retail,* Construction, Food

San Diego (433)
B. *Administration*
C. Transp. equipment, *Retail*

Oceanside (13)
B. *Administration*
C. *Retail, Eating places,* Construction

Newport Beach (11)
C. Construction, *Retail, Finance, Eating places*

Costa Mesa (12)
B. Construction

Santa Ana (46)
C. *Retail,* Construction, *Administration*

Orange (10)
C. Construction, *Medical, Wholesale*

Anaheim (15)
C. *Wholesale,* Food

Fullerton (14)
C. Food, *Education*

Corona (10)
B. *Wholesale*
C. Construction

Riverside (47)
C. *Administration, Retail, Wholesale*

San Bernardino (136)
C. *Railroads, Administration,* Construction

Ontario (23)
C. El. machinery, Prim. metal, Food, Construct.

Pomona (35)
C. *Wholesale,* Construct., *Retail*

Los Angeles (3,997)

Oxnard (22)
A. *Administration*
C. *Wholesale*

San Buenaventura (17)
B. Mining
C. *Administration, Retail*

Santa Paula (11)
B. *Wholesale*
C. Construction, *Administration,* Mining

Santa Barbara (45)
C. *Retail, Households,* Construction, *Medical*

Santa Maria (10)
C. *Retail, Wholesale*

San Luis Obispo (14)
C. *Education, Railroads, Retail,* Construction, *Administration*

Bakersfield (102)
C. Mining, *Retail,* Construction

Tulare (12)
B. *Retail*
C. *Education, Wholesale*

Visalia (12)
B. *Retail*
C. *Education, Administration*

Hanford (10)
B. *Retail*

Fresno (131)
C. *Retail, Wholesale,* Construction

Madera (10)
C. Construction, *Retail, Education, Administration*

Merced (15)
B. *Retail*
C. *Administration,* Construction, *Eating places*

Modesto (17)
C. *Retail,* Food

Stockton (113)
C. *Administration, Retail*

Lodi (14)
B. Food
C. *Retail,* Construction

Sacramento (212)
B. *Administration*
C. *Railroads,* Construct.

Monterey-Seaside (26)
B. Construction
C. Food, *Eating places, Administration*

Salinas-Alisal (31)
B. *Wholesale*
C. *Retail,* Construction, Food

Watsonville (12)
C. *Wholesale,* Food, *Retail, Eating places,* Construction

Santa Cruz (22)
C. *Retail,* Food

Santa Jose (166)
C. Food, Construction

San Francisco-Oakland (2,022)
C. *Administration*

Pittsburg (13)
A. Primary metal
C. Fabricated metal

Antioch (11)
A. Nondurable
B. Primary metal
C. Chemicals

Napa (14)
B. Transp. equipment
C. *Medical, Retail*

Petaluma (10)
C. Food, *Wholesale, Retail*

Santa Rosa (18)
C. *Retail, Medical, Wholesale*

Chico (12)
C. *Retail, Education*

Redding (10)
C. *Retail,* Construction, Furniture & Lumber

Eureka (23)
A. Furniture & Lumber

# APPENDIX 2. The Industrial Structure of Cities in The United States with More Than 1,000,000 Inhabitants*

| Industries | NY | Chi | LA | Phi | Det | Bos | SF | Pitt | SL | Cle | Wash | Balt |
|---|---|---|---|---|---|---|---|---|---|---|---|---|
| Mining . . . . . . . | 0.1 | 0.1 | 0.5 | 0.1 | 0.1 | 0.0 | 0.1 | 0.8 | 0.2 | 0.1 | 0.0 | 0.1 |
| Construction . . . . . | 5.1 | 4.4 | 7.6 | 6.0 | 4.7 | 5.8 | 7.1 | 5.4 | 5.0 | 5.1 | 6.6 | 6.6 |
| Furniture & Lumber. . | 0.7 | 1.2 | 1.4 | 0.8 | 0.4 | 0.7 | 0.8 | 0.4 | 0.9 | 0.8 | 0.2 | 1.0 |
| Primary metal . . . . | 0.7 | 5.4 | 1.0 | 1.4 | 2.9 | 0.5 | 1.0 | 18.5 | 2.8 | 6.8 | 0.1 | 6.6 |
| Fabricated metal . . . | 1.5 | 3.6 | 2.0 | 2.4 | 2.7 | 1.5 | 1.7 | 3.1 | 2.6 | 4.7 | 1.3 | 2.7 |
| Machinery . . . . . | 1.9 | 4.4 | 1.9 | 2.5 | 4.5 | 2.3 | 1.5 | 1.6 | 2.5 | 6.7 | 0.2 | 1.0 |
| Electrical machinery . . | 2.0 | 4.7 | 1.0 | 3.4 | 0.7 | 3.8 | 0.7 | 4.0 | 2.9 | 3.4 | 0.2 | 1.4 |
| Motor vehicles . . . . | 0.5 | 0.8 | 1.0 | 1.0 | 28.0 | 0.5 | 0.9 | 0.3 | 1.8 | 4.5 | 0.1 | 1.1 |
| Transport. equipment . | 1.1 | 0.8 | 4.7 | 2.0 | 0.3 | 1.6 | 2.0 | 0.7 | 1.1 | 1.8 | 0.1 | 3.0 |
| Oth. durable goods . . | 3.4 | 2.9 | 2.2 | 2.0 | 1.3 | 1.9 | 1.2 | 2.6 | 2.1 | 1.9 | 0.3 | 1.7 |
| Food manufacturing . . | 2.3 | 4.5 | 2.4 | 3.2 | 1.8 | 2.9 | 3.3 | 2.8 | 5.3 | 1.8 | 0.9 | 3.6 |
| Textile manufacture . . | 1.8 | 0.4 | 0.3 | 4.1 | 0.1 | 1.2 | 0.3 | 0.2 | 0.4 | 0.9 | 0.0 | 0.6 |
| Apparel manufacture . | 7.4 | 1.6 | 2.5 | 4.0 | 0.3 | 2.5 | 0.9 | 0.2 | 2.7 | 1.9 | 0.0 | 2.8 |
| Printing . . . . . . . | 2.8 | 3.5 | 1.9 | 2.7 | 1.5 | 2.3 | 2.3 | 1.4 | 2.1 | 2.3 | 4.0 | 2.0 |
| Chemicals . . . . . . | 2.0 | 1.5 | 0.9 | 1.7 | 1.4 | 1.2 | 1.1 | 1.0 | 2.4 | 1.9 | 0.1 | 1.9 |
| Other nondurable . . . | 2.9 | 2.6 | 2.8 | 4.6 | 1.3 | 6.0 | 2.1 | 1.4 | 4.4 | 1.6 | 0.2 | 2.0 |
| Railroads. . . . . . . | 1.5 | 3.8 | 1.4 | 2.1 | 1.2 | 1.7 | 2.5 | 4.3 | 4.3 | 2.7 | 1.8 | 2.7 |
| Trucking. . . . . . . | 1.3 | 1.7 | 1.2 | 1.3 | 1.2 | 1.3 | 1.4 | 1.2 | 1.9 | 1.5 | 0.7 | 1.4 |
| Other transportation . . | 3.4 | 2.0 | 1.7 | 2.4 | 1.5 | 2.5 | 4.1 | 1.4 | 1.5 | 1.8 | 2.5 | 3.3 |
| Telecommunications . . | 1.7 | 1.6 | 1.7 | 1.3 | 1.2 | 1.7 | 2.0 | 1.4 | 1.4 | 1.3 | 1.5 | 1.2 |
| Utilities . . . . . . . | 1.6 | 1.3 | 1.6 | 1.4 | 1.7 | 1.6 | 1.4 | 1.7 | 1.7 | 1.6 | 1.4 | 1.9 |
| Wholesale trade . . . . | 5.5 | 4.4 | 5.1 | 3.9 | 3.1 | 4.9 | 5.6 | 3.8 | 5.2 | 4.0 | 2.2 | 4.1 |
| Food stores. . . . . . | 3.7 | 3.1 | 3.2 | 3.7 | 3.2 | 3.6 | 3.2 | 3.9 | 3.1 | 3.3 | 2.3 | 3.9 |
| Eating places . . . . . | 3.6 | 3.4 | 3.9 | 3.4 | 3.0 | 3.5 | 4.3 | 2.9 | 3.3 | 3.0 | 3.3 | 3.6 |
| Other retail trade . . . | 9.2 | 10.0 | 11.6 | 10.1 | 9.2 | 10.7 | 10.8 | 9.9 | 10.1 | 9.3 | 9.7 | 10.4 |
| Finance . . . . . . . | 7.1 | 4.9 | 5.4 | 4.8 | 3.5 | 5.9 | 6.7 | 3.9 | 4.5 | 3.8 | 4.9 | 4.4 |
| Business service . . . . | 1.9 | 1.6 | 1.7 | 1.2 | 1.2 | 1.3 | 1.5 | 1.0 | 1.1 | 1.3 | 1.0 | 0.9 |
| Repair services . . . . | 1.4 | 1.2 | 2.0 | 1.4 | 1.3 | 1.5 | 1.7 | 1.2 | 1.3 | 1.4 | 1.2 | 1.3 |
| Private households . . | 2.9 | 1.7 | 3.2 | 3.2 | 1.8 | 2.3 | 2.9 | 2.0 | 2.7 | 2.2 | 4.4 | 4.3 |
| Hotels . . . . . . . . | 1.1 | 1.2 | 1.0 | 0.6 | 0.7 | 1.0 | 1.4 | 0.7 | 1.0 | 0.8 | 1.5 | 0.6 |
| Oth. personal services . | 2.8 | 2.7 | 3.0 | 2.8 | 2.4 | 2.8 | 3.0 | 2.4 | 2.7 | 2.0 | 2.9 | 2.9 |
| Entertainment . . . . | 1.4 | 1.0 | 3.2 | 0.9 | 1.0 | 1.0 | 1.3 | 1.1 | 1.0 | 1.1 | 1.0 | 1.1 |
| Medical services . . . | 3.4 | 2.7 | 3.8 | 3.3 | 2.7 | 4.7 | 4.2 | 3.3 | 3.3 | 3.3 | 4.0 | 3.8 |
| Education . . . . . . | 2.9 | 2.8 | 3.9 | 3.1 | 2.7 | 4.4 | 4.2 | 3.2 | 2.8 | 3.0 | 3.9 | 3.0 |
| Oth. professional serv. . | 2.6 | 2.1 | 2.3 | 2.0 | 1.6 | 2.5 | 2.4 | 2.2 | 2.0 | 2.1 | 3.6 | 1.6 |
| Public administr. . . . | 4.5 | 4.0 | 4.8 | 4.7 | 3.4 | 6.2 | 8.2 | 3.8 | 5.4 | 4.0 | 32.1 | 5.8 |
| | 100.0 | 100.0 | 100.0 | 100.0 | 100.0 | 100.0 | 100.0 | 100.0 | 100.0 | 100.0 | 100.0 | 100.0 |

* NY New York, Chi Chicago, LA Los Angeles, Phi Philadelphia, Det Detroit, Bos Boston, SF San Francisco, Pitt Pittsburgh, SL St. Louis, Cle Cleveland, Wash Washington, Balt Baltimore.

Printed and bound by CPI Group (UK) Ltd, Croydon, CR0 4YY

22/10/2024

01777647-0011